MINECRAFT

MATHS
OFFICIAL WORKBOOK
AGES 10-11

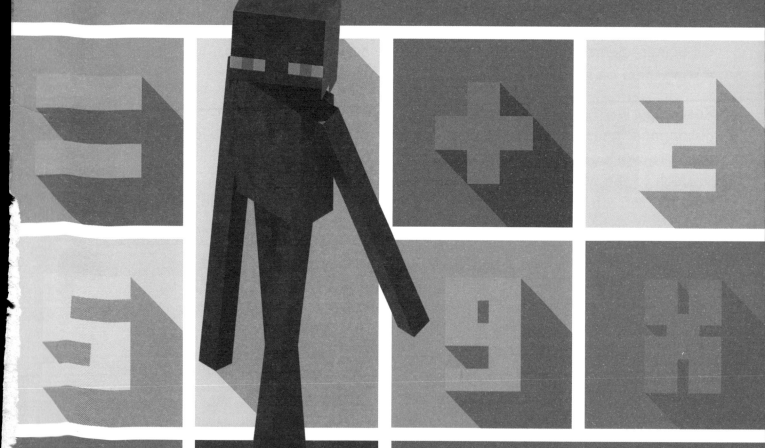

DAN LIPSCOMBE AND
KATHERINE PATE

INTRODUCTION

HOW TO USE THIS BOOK

Welcome to an exciting educational experience! Your child will go on a series of adventures through the amazing world of Minecraft, improving their maths skills along the way. Matched to the National Curriculum for maths for ages 10–11 (Year 6), this workbook takes your child into fascinating landscapes where our heroes Isaac and Rosa embark on building projects and daring treasure hunts…all while keeping those pesky mobs at bay!

As each adventure unfolds, your child will complete topic-based questions worth a certain number of emeralds . These can then be 'traded in' on the final page. The more challenging questions are marked with this icon to stretch your child's learning. Answers are included at the back of the book.

MEET OUR HEROES

Isaac enjoys learning, a lot. Whether it is science or geography, he will gobble up information. Rather than at home, he is often found exploring new places. He likes nothing more than wandering home with an inventory full of ingredients and spending the rest of the day brewing potions and testing them on his friends.

Rosa experiments with machines and redstone. If ever a minecart track or piston machine is needed, Rosa will know how to make it work. She keeps trying to work out how Nether portals work, but always ends up pulled through to the other dimension! Rosa loves slimes because slimeballs are used to craft sticky pistons. If she could keep one as a pet, she would!

First published in 2021 by Collins
An imprint of HarperCollins*Publishers*
1 London Bridge Street, London, SE1 9GF

HarperCollins*Publishers*
1st Floor, Watermarque Building, Ringsend Road, Dublin 4, Ireland

Publisher: Fiona McGlade
Authors: Dan Lipscombe and Katherine Pate
Project management: Richard Toms
Design: Ian Wrigley and (cover) Sarah Duxbury and Keith Moore
Typesetting: Nicola Lancashire at Rose and Thorn Creative Services

Special thanks to Alex Wiltshire, Sherin Kwan and Marie-Louise Bengtsson at Mojang and the team at Farshore

Production: Karen Nulty

ISBN: 978-0-00-846279-6
British Library Cataloguing in Publication Data.
A CIP record of this book is available from the British Library.
2 3 4 5 6 7 8 9 10
Printed in the United Kingdom

MIX
Paper from responsible source
FSC
www.fsc.org
FSC™ C007454

This book is produced from independently certified FSC™ paper to ensure responsible forest management.

For more information visit: www.harpercollins.co.uk/green

CONTENTS

NUMBER AND PLACE VALUE

TOUGH IN THE TUNDRA

Much like the plains, the snowy tundra is mostly large and flat. However, the tundra is a tougher place to live. Snow blankets every surface in sight. The bright white stretches in every direction, broken up occasionally by the odd tree. Rabbits zip about at the sound of footsteps and the polar bears camouflage with the landscape.

LOTS TO FIND

It can be incredibly quiet in this biome. With so few animals, it is a peaceful but chilly place to explore. Any explorer who brings a shovel can dig up snowballs to throw or form into blocks. Those with pickaxes enchanted with Silk Touch can harvest the ice from frozen rivers.

SPARSE SURROUNDINGS

Despite the lack of food, villagers still throw caution to the wind and build bustling settlements in the cold. These villages are usually built from spruce wood obtained from the surrounding trees.

VILLAGE VISIT

After walking for an awfully long time, through snow and over frozen rivers, Isaac finds a small village to stop and rest. There are some shops, empty houses where travellers can rest and even a library.

PLACE VALUE AND ROUNDING

Isaac's journey to the snowy tundra has taken a very long time. Looking at his map, he ponders how many blocks he walked...that gets him thinking about some very big numbers!

 1

a)   Write in digits:

three million, six hundred and eleven thousand, two hundred

....

b)   Write in words: 2,809,000

....

2

Isaac has travelled a lot over the last year. Complete this table by rounding the numbers of blocks travelled across different continents.

Blocks travelled	Rounded to nearest 1,000	Rounded to nearest 10,000	Rounded to nearest 1,000,000
4,092,781			
6,935,092			
8,249,711			

3

Write **<** or **>** in the box to make each statement correct.

a)   126.2 ☐ 162.09

b)   166,434 ☐ 163,343

c)   5.564 ☐ 5.654

d)   257,979 ☐ 259,779

Isaac can take a rest in one of the small, empty houses available in the village. While he relaxes, continue with the questions.

4

What is the value of the digit 3 in each number?

a) 52.83 ....

b) 235,419 ....

c) 7.943 ....

5

Colour the box with a number that has a digit 4 with a value of four hundred thousand.

| 4,005,812 | 940,027 | 4,005 | 1,340,087 | 1,491,226 |

6

Write the numbers on the blocks in order, from largest to smallest.

32.6   326.2   3.262   3.662   3.26

7

Complete this table by rounding the numbers.

	Rounded to 1 decimal place	Rounded to the nearest whole number
4.25		
7.41		
9.82		

8

 Circle the number closest in value to 0.09

0.951

0.95

0.099

0.010

COLOUR IN HOW MANY EMERALDS YOU EARNED

NEGATIVE NUMBERS

The snow makes everything cold. Isaac is not used to the cold weather as his house is in the plains, where it is much warmer.

1

Look at this thermometer.

a) What is the temperature shown on the thermometer? ....°C

b) The temperature decreases by 8°C.

What is the new temperature? ....°C

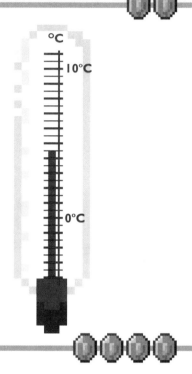

2

a) Write down the temperature shown on each thermometer.

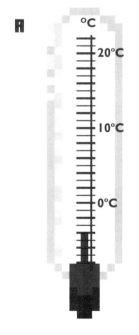 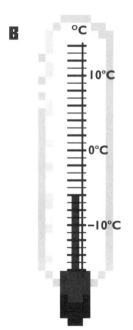 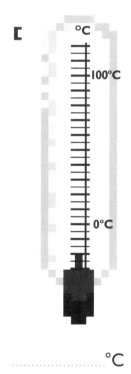

.......................°C   ....°C   ....°C

b) Which thermometer shows the coldest temperature? ....

The cold does not seem to bother the villagers because they have grown up with these conditions. They are used to temperatures below 0°C.

3

The temperature during the day is 6°C.
By midnight, the temperature has fallen by 8°C.

What is the temperature at midnight? ....°C

4

Work out:

a) 6 – 9 = ...

b) 2 – 9 = ...

c) 7 – 10 = ...

d) –2 + 4 = ...

e) –4 + 9 = ...

f) –5 + 7 = ...

5

Draw a line to join the two boxes that have numbers with a difference of 15.

| –10 |

| –7 |

| 3 |

| –4 |

| 8 |

| 15 |

MENTAL CALCULATIONS

When crafting, building or trading, Isaac needs to carry out maths calculations in his head.

Show him how to do mental calculations with numbers big and small.

1

Try to calculate these mentally, or use a suitable written method.

a) 275,843 + twenty thousand = ...

b) 411,607 − forty thousand = ...

c) 3,728,910 − two hundred thousand = ...

d) 7,390,190 + six hundred thousand = ...

2

Work out:

a) 38 + 39 = ...

b) 10,000 − 1 = ...

c) 50 + 70 = ...

d) 130 − 90 = ...

e) 199 − 50 = ...

f) 199 + 30 = ...

g) 8,334 − 1,300 = ...

h) 1,600 + 145 = ...

Isaac visits a shop to buy supplies. He only has a small amount of money, so he must look carefully at the price list before buying.

3

Here are the prices in pounds and pence of three items in the shop.

| £1.99 | 65p | £2.05 |

a) How much do these three items cost in total? £ ....

b) Isaac buys one of each item.

How much change does he get from £10? £ ....

4

Work out:

a) 230 + 60 + 180 = ...

b) 550 − 60 + 170 = ...

c) 90,000 + 900 − 9 = ...

d) 5,200 + 52,000 = ...

e) 2,497 + 2,503 = ...

f) 4,001 − 302 = ...

Isaac collected lots of coal on his travels. He doesn't need it all so he donates some of it to the villagers.

5

 Isaac has 299 coal.

He gives the village weaponsmith 61 coal and the armourer 89 coal.

How much coal does Isaac have left? .... coal

COLOUR IN HOW MANY EMERALDS YOU EARNED

ADDITION AND SUBTRACTION

The villagers seem grateful for Isaac's generous donation of coal. They lead him to the side of a mountain. He is puzzled until he sees a door. Inside are hundreds of chests: there must be thousands of items! While he takes it all in, answer these questions which include some equally big numbers.

1

Complete these addition and subtraction calculations.

a)
```
   1 5 2 3 6
 + 2 0 3 7 4
 _____
```

b)
```
   3 4 7 . 2 5
 + 1 0 6 . 4 7
 _____
```

c)
```
   2 6 5 1 . 4
 - 1 9 3 7 . 2
 _____
```

d)
```
   9 8 7 3 2 6
 - 3 4 6 9 8 1
 _____
```

2

Complete these addition and subtraction calculations.

a)
```
   3 7 4 8 9 1
 +     1 2 5 7 . 6
 _____
```

b)
```
   3 2 7 6 . 5
 +    1 4 . 3 9
 _____
```

c)
```
   5 7 2 . 7 4
 -    3 7 . 5
 _____
```

d)
```
   5 0 0
 - 2 1 1 . 8 6
 _____
```

Isaac has never seen so many chests before! This massive storage facility must have been created by explorers who were here in the past. It now seems to be used by the villagers and to support any adventurers, like Isaac, who pass by.

3

Work out:

a) 785 + 10,365 = ...

b) 9,632,470 − 5,672 = ...

c) 30.72 + 456.3 = ...

d) 5,000 − 158.46 = ...

e) .... + 57.6 = 2,000

f) 47.253 − .... = 19.5

4

Each box holds the total of the numbers in the two boxes below it.

Work out the numbers in the empty boxes.

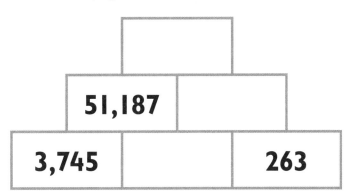

51,187

3,745   263

5

Complete these calculations which have missing digits.

a)
```
    5 4 6 2 □
  −   2 □ 8 3 1
  _____
    3 0 □ 9 0
```

b)
```
    2 9 7 □ 0
  +   6 □ 4 1 6
  _____
    9 3 □ 7 6
```

ESTIMATION AND INVERSE CALCULATIONS

Isaac is not sure what to take from the store. Each chest is labelled, and he spends some time walking among the rows. He is speechless at how much the village has stored. It would be almost impossible to count all the items kept in the chests, but he could make some estimates.

1

Estimate the answer to each calculation by rounding each number to the nearest thousand.

a) 5,617 + 2,704 − 4,984 = ...

b) 7,380 + 1,476 + 3,791 = ...

c) 9,128 − 3,467 − 3,423 = ...

d) 6,766 − 4,951 − 1,003 = ...

2

Write an inverse calculation for each given calculation.

a) 53,872 + 23,782 = 77,654 ....

b) 60,936 − 37,058 = 23,878 ....

c) 32,611 + 18,543 = 51,154 ....

d) 93,137 − 47,675 = 45,462 ....

3

The mountain storage has 17,826 items. A few years ago it only had 6,902 items.

Isaac thinks the difference of these numbers is 11,128.

Write an inverse calculation to show that Isaac is wrong.

4

Estimate the answer to each calculation by rounding each number to the nearest hundred thousand.

a)   749,613 + 537,229 =

b)   1,510,717 – 391,426 =

c)   3,423,812 + 865,106 + 234,052 =

d)   2,557,328 – 577,426 – 437,932 =

5

 Calculate an estimate for this calculation, to the nearest whole number.

$$43.276 - 19.57 + 3.8 - 9.45$$

COLOUR IN HOW MANY EMERALDS YOU EARNED

NUMBER PROBLEMS

Taking inspiration from the village, Isaac wants to build a museum when he returns home. It will have rooms themed on each biome, including a special feature on the Nether. Isaac buys a book and quill from the local shop and jots down his ideas.

1

Isaac wants to have 20,000 exhibits in his museum.

He already has 12,250 exhibits.

He will collect 2,750 more before he opens the museum.

His friend Rosa will give him 1,800 exhibits.

How many more exhibits does Isaac need to reach his 20,000 target? ....

2

Museum visitors will score points for finding hidden dandelions and poppies.

They will win a prize if they score 10,000 points.

One villager scores 4,845 points for dandelions and 5,177 points for poppies.

Does the villager win a prize? Show your working out.

....

3

Isaac will give out some free tickets when the museum opens.

This table shows the numbers of free tickets for each day.

	Wed	Thu	Fri	Sat	Sun	Mon
Children	2,516	2,457	1,084	5,258	6,041	987
Adults	1,369	946	843	3,217	4,279	726

At the weekend, how many more children than adults get free tickets?

....

It is Isaac's last day in the village. He hears something outside. Thousands of adventurers are passing by on an organised trek.

 4

14,814 adventurers signed up for the trek.

1,893 adventurers did not turn up to do the trek.

735 of the adventurers who started the trek did not finish it.

How many adventurers finished the trek? ....

The sun has started to set. Isaac notices a zombie shuffling along at the edge of the village. Running to the village bell, he begins ringing it to warn the villagers of the danger.

 5

These times show how fast each villager dashed to the safety of their home.

1	2	3	4	5	6

55.75 seconds	49.41 seconds	58.59 seconds	46.25 seconds	51.47 seconds	53.34 seconds

a)   What was the total time taken by villagers 1 and 5?   .... seconds

b)   How much faster was villager 4 than villager 3?   .... seconds

c)   Two of the villagers took a total time of 108 seconds.

Which two villagers are they?   .... and ....

d)   Calculate the total time taken by villagers 4 and 6.

How much quicker was their total time than
that of villagers 1 and 5?   .... seconds

Isaac waves goodbye as he leaves the snowy tundra village. Looking at the map, he plots a course for home. He will travel through several biomes on the way.

 6

This table shows the villager population of four biomes 10 years ago and now.

Biome	Plains	Savanna	Jungle	Swamp
10 years ago	12,668,830	5,095,969	2,596,850	1,370,921
Now	19,138,831	5,062,416	2,910,228	1,685,267

a) Which biome's population has increased
the most over the 10-year period?   ....

b) How many villagers now live in the plains,
savanna, jungle and swamp altogether?   ....

 7

A nearby jungle biome has 138,578 iron ore and 119,843 copper ore beneath the ground.

In one month 96,775 ore is mined.

The next month 100,000 ore is mined.

How much iron and copper ore is left in this biome now?

....

8

💜 A fundraising target is £10,000.

The amounts raised so far are:

£500   **£1,250**   **£59.75**   **£312.40**   **£5,226.04**

How much more money needs to be raised to reach the target?

£ ....

ADVENTURE ROUND-UP

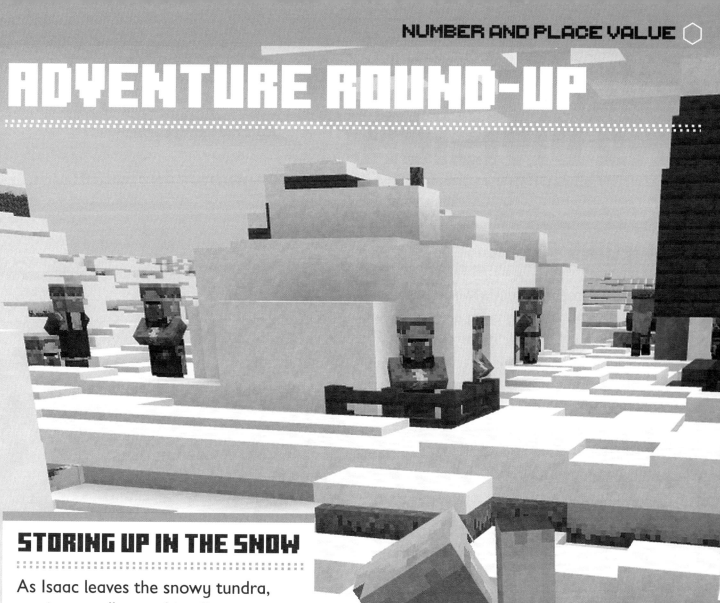

STORING UP IN THE SNOW

As Isaac leaves the snowy tundra, waving goodbye to his villager friends, he reflects on their lives there. In very cold places it is difficult to grow crops or raise animals. As a result, the villagers had made use of a huge store where they could keep supplies for times of need.

BOOK WITH A BONUS

Before leaving, Isaac took one enchanted book from the village store. The book will be particularly important for his continuing journey, as it will enchant his pickaxe with Fortune III. This means that whenever he harvests materials, they will drop more than usual. Imagine how many diamonds he can collect now!

MULTIPLICATION AND DIVISION

TOWERING TREES

A tall birch forest is exactly what you might think it is. It is a grassy biome similar to the plains, with towering birch trees growing close together. Flowers dot the forest floor and animals weave in and out of the black and white trunks.

SECRET STRONGHOLD

It is what lies below the ground that makes this tall birch forest special. Underneath layers of dirt and stone is a maze of tunnels and stairs, rooms sealed off by iron doors and others filled with books or even mobs. This is called a stronghold.

ENDER DRAGON ENCOUNTER?

At the centre of the stronghold is a portal built into the floor. With eyes of Ender placed in empty slots, an adventurer can spark the portal to life and dive into a whole new realm. By stepping through, a hero can come face to face with the Ender Dragon. Whether in search of riches or glory, adventurers stand before the portal wondering if they are ready for that ultimate challenge.

STEPPING OUT

Rosa steps out of her house with her inventory half full of helpful items, tools and weapons. She has spent ages working up to this: fighting mobs, crafting armour and swords, and enchanting everything to make her safer and stronger. With a stack of eyes of Ender, off she goes into the tall birch forest.

FACTORS, MULTIPLES, POWERS AND PRIMES

Rosa knows how to find the End portal, which sits within a stronghold. She is prepared, yet nervous. To find the right direction, she must throw an eye of Ender into the air. It will drift in a particular direction, before falling to the ground. That's the way she needs to walk.

1

Rosa may need to throw the eye of Ender many times before getting close to the stronghold.

Circle:

a) the multiples of 20:

10   60   110   150   180   300   470

b) the multiples of 25:

5   10   50   100   165   205   350

2

Find the lowest common multiple of:

a) 3 and 6   ....

b) 12 and 15   ....

c) 20 and 25   ....

3

a) What are the common factors of 24 and 32?   ....

b) What is the lowest common multiple of 24 and 32?   ....

21

Rosa thinks she has found the stronghold. It seems she is standing above it. The eye of Ender floats up in a straight line and flutters to the ground. The stronghold could be 2 layers below, or 22 layers! It is never safe to dig straight down. There could be lava or a big drop! Rosa will dig a larger square, removing one layer at a time. Help her to dig safely by answering these questions.

4

Write **<**, **>** or **=** in the boxes to make each statement correct.

a) 4^2 ☐ 2^2 b) 8^2 ☐ 4^3 c) 3^3 ☐ 5^2

5

Look at these number cards:

| 1 | 2 | 3 | 4 | 5 | 6 |

Use each of these number cards once only to make three 2-digit numbers that complete the statements.

☐☐ is a prime number

☐☐ is a common multiple of 8 and 28

☐☐ is a common factor of 64 and 96

As Rosa digs the final layer of dirt away, she uncovers a new material – stone bricks. This must be the ceiling of the stronghold. Harvesting these stones, she drops down into a corridor. Now she must find her way to the portal. Answer each question to help her move in the right direction.

6

Work out:

a) $10^3 + 9^2 - 5^2 =$ ....

b) $7^2 - 2^3 + 6^2 =$ ....

7

Find a 2-digit prime number which is still a prime number when the digits are reversed.   ....

8

Write a number less than 30 in each box to make the statement correct.

Prime number		Square number		Square number
☐	+	☐	=	☐

MULTIPLYING AND DIVIDING MENTALLY

0.07×4

$$0.07 = \frac{7}{100} \text{ or } 7 \div 100$$

$$0.07 \times 4 = 7 \div 100 \times 4$$
$$= 7 \times 4 \div 100$$
$$= 28 \div 100$$
$$= 0.28$$

After only a few steps, Rosa starts to hear skeletons approaching. An arrow whizzes past her head. Two skeletons stand between Rosa and the next door but she defeats them with her super sword skills.

1

Work out:

a) $245 \div 10 =$ ....

b) $51 \times 10 =$ ....

c) $156 \div 100 =$ ....

d) $16 \times 100 =$ ....

e) $810 \div 9 =$ ....

f) $50 \times 30 =$ ....

g) $180 \div 3 =$ ....

h) $2{,}400 \div 6 =$ ....

2

Work out:

a) $5.1 \div 10 =$ ....

b) $5.98 \times 100 =$ ....

c) $261.9 \div 100 =$ ....

d) $9.063 \times 10 =$ ....

e) $2.07 \div 10 =$ ....

f) $0.604 \times 1{,}000 =$ ....

g) $431 \div 1{,}000 =$ ....

h) $0.078 \times 100 =$ ....

Rosa's path to the End portal is complicated. Some corridors lead to dead-ends, while others go around and around until they meet up again. Occasionally she must use her pickaxe to break down a wall.

3

Fill in the missing numbers.

a) .... × 100 = 560

b) 1,356 ÷ 100 = ...

c) .... ÷ 10 = 34

d) .... ÷ 1,000 = 7.291

e) 100 × .... = 47.8

f) .... ÷ 100 = 247.3

4

Rosa chips away one piece of stone with mass 0.5 kg.

What is the total mass of 7 of these pieces of stone?

.... kg

5

Complete these calculations.

a) 0.6 × 7 = ...

b) 0.9 × 4 = ...

c) 0.3 × 8 = ...

d) 0.05 × 4 = ...

e) 0.09 × 8 = ...

f) 0.07 × 8 = ...

g) 0.7 × .... = 8.4

h) 0.06 × .... = 0.54

COLOUR IN HOW MANY EMERALDS YOU EARNED

WRITTEN MULTIPLICATION

Rosa finds herself in a corridor with an iron door at the end. Approaching the door, she presses the button to the side and the door swings open to reveal an amazing library! There must be hundreds of books. What a strange place for a library!

1

There are 28 sections in the library.

Each section holds 36 books.

How many books are there in total? .... books

2

Rosa tries to work out the total number of pages in the books from one bookcase.

She starts filling in this multiplication grid for her calculation.

×	400		3
	12,000		
6		420	

Fill in the empty boxes and calculate the total.

Total: ....

3

Use long multiplication to help Rosa with some more calculations.

a)
```
      5 2 6
  ×     3 2
  _____

  _____

  _____
```

b)
```
    1 4 7 8
  ×     2 6
  _____

  _____

  _____
```

c)
```
    4 5 . 2 6
  ×       2 4
  _____

  _____

  _____
```

Rosa exits the library using a different door, hoping that she is getting somewhere helpful. This stronghold is huge! It must have taken someone, or something, ages to build. It's like a massive maze.

4

In pounds and pence, a stack of stone bricks costs £7.32

How much do 25 stacks of these stone bricks cost?

Stone bricks

£7.32 per stack

£ ....

5

Use long multiplication to calculate 6,847 × 74

....

6

Rosa does some more calculations to try to figure out the size of the stronghold.

Find the missing numbers in her calculations.

a)
```
      5 □ 6 3
×       □ 6
    3 2 7 7 8
  1 6 3 8 □ 0
  1 9 6 6 6 8
```

b)
```
      3 1 □ □
×         2 7
    2 2 1 6 □
  6 3 3 4 0
  8 5 5 0 9
```

WRITTEN DIVISION

Rosa must admit it. She is lost. She takes several turns in the corridors and they lead to nothing. She keeps on walking, even though her hunger is getting low. Help to maintain her focus by answering these questions.

1

Work out the answers to these division calculations.

a) $8 \overline{)136}$

b) $14 \overline{)252}$

c) $13 \overline{)572}$

d) $16 \overline{)192}$

e) $25 \overline{)4250}$

f) $5 \overline{)71.5}$

2

How Rosa wishes she had a stack of apples to give her a boost!

In pounds and pence, a stack of 15 apples costs £4.05

How much does one apple cost? ....P

3

Work out:

a) 6,230 ÷ 7 = ...

b) 7,578 ÷ 18 = ...

c) 4,715 ÷ 23 = ...

d) 3.99 ÷ 21 = ...

4

Rosa can stack 64 stone bricks in one storage slot of her inventory.

How many storage slots would she need
available to take 1,020 stone bricks?

.... storage slots

5

The creepy stronghold has 1,680 mossy stone bricks.

The mossy stone bricks are equally spread across 24 rooms.

How many mossy stone bricks are in each room?

.... mossy stone bricks

DIVISION WITH REMAINDERS

Suddenly, turning a corner and expecting yet another dead-end, Rosa finds a staircase with two doors at the top. These doors open to reveal the End portal.

 1

Rosa has 188 eyes of Ender.

These eyes of Ender can be used to fill portals.

One portal holds 12 eyes of Ender in total.

a) How many portals could Rosa fill? ....

b) How many eyes of Ender would be left over? ....

 2

While Rosa prepares her eyes of Ender, use division to work out the answers to these. Give remainders as decimals.

a) $5 \overline{)193}$

b) $12 \overline{)495}$

c) $24 \overline{)372}$

d) $4 \overline{)423}$

e) $16 \overline{)2050}$

f) $8 \overline{)193.2}$

Rosa places her eyes of Ender into the empty slots around the frame of the portal. As the last one is secured, the portal opens. Rosa finds herself staring into a dark sky full of twinkling stars. It is tempting to jump in, but she knows the Ender Dragon awaits on the other side. She must prepare. Instead of trudging back through the maze, Rosa digs a staircase back up to the surface. Answer these questions while she digs.

3

A number divided by 45 is 156 r32.

What is the number? ....

4

The stronghold has some empty prison cells.

Each cell can hold 8 prisoners.

What is the minimum number of cells
needed for 146 prisoners? ....

5

Use division to work out the answers. Give remainders as fractions.

a) 6)93

b) 9)327

c) 12)195

d) 14)583

e) 22)1395

f) 25)3055

COLOUR IN HOW MANY
EMERALDS YOU EARNED

MULTIPLICATION AND DIVISION PROBLEMS

Rosa pops out of the ground. She has left a stairway tunnel behind her which will lead right back to the portal. However, she is in the middle of nowhere, so must mark the entrance somehow. She decides a tower lit with torches is best, as it can be seen from far away. Once this is built, she begins the walk home and plans the next stage. Answer these questions while she sets off.

 1

Rosa has heard of mysterious ships being moored off the islands of the End.

She imagines that there are 9 ships that can each carry 1,368 passengers.

Each ship sails 19 times a year.

What is the maximum number of passengers
that can be carried in total in one year?

.... passengers

2

Rosa is thinking about what she'll need to do to prepare for her journey to the End. She buys 35 stacks of carrot seeds. Each stack holds 64 seeds.

Rosa later loses 800 of the seeds. She then loses another 800.

How many seeds does she have left?   .... carrot seeds

3

Complete these calculations.

a)   .... ÷ 28 = 98

b)   32 × .... = 472

c)   4,770 ÷ .... = 265

d)   .... ÷ 37 = 86 r23

As Rosa walks home, she occasionally stops to chop down birch trees. She might as well harvest and fill her inventory on the journey. She'll drop in to see Isaac, too.

4

Here are the prices of some young trees in pounds and pence:

Oak tree	Birch tree	Acacia tree

£4.50 each	**£7 each**	**£3.50 each**

a) What is the total cost of 37 oak trees and 15 acacia trees?   £...................

b) A forester has £357 to spend on birch trees and acacia trees.

He buys 23 birch trees.

How many acacia trees can he buy?   .... acacia trees

c) New trees are delivered in lorry loads of 400.

How many lorry loads are needed to deliver 9,600 trees?   .... lorry loads

5

 There are 12 areas of birch trees.

There are 50 birch trees in each area.

28 birch trees are cut down each week for 2 weeks.

Then 29 birch trees are cut down each week for 6 weeks.

How many birch trees are left?   .... trees

COLOUR IN HOW MANY EMERALDS YOU EARNED

ORDER OF OPERATIONS

The order of operations is:

Brackets   Indices   Division and Multiplication   Addition and Subtraction
(squares or cubes)

By the time Rosa reaches Isaac's house, she has calculated all the materials she will need to fight the Ender Dragon. All these materials will be used to make weapons, craft potions, enchant weapons and construct strong armour.

1

Work out:

a) $50 - 10 + 30 - 40 =$ ....

b) $12 - 20 + 6 =$ ....

c) $60 - 40 \div 2 =$ ....

d) $80 + 10 \times 10 =$ ....

e) $20 + 5 \times 4 - 4 =$ ....

f) $2 + 3 \times 6 =$ ....

2

Work out:

a) $(20 + 2) \times 5 =$ ....

b) $4 + 3^2 - 5 =$ ....

c) $200 \times 100 \div 10 \times 2 =$ ....

d) $1{,}000 \div 10 + 500 \times 102 =$ ....

e) $(3 + 4) \times 2 + 4 - 2 + 5 =$ ....

f) $7 - 3 + 4 \times 2^3 =$ ....

3

 Fill in the missing numbers to make these statements correct.

a) $3 \times 8 + 12 =$ .... $^2 \times 10 - 4$

b) $5 +$ .... $\times 6 \div 3 = 16 \div 4 \times 2 + 5$

c) $3^3 - 22 + 3 =$ .... $^3 - 56$

COLOUR IN HOW MANY EMERALDS YOU EARNED

ADVENTURE ROUND-UP

ROSA RUSHES IN

Rosa thumps on Isaac's door. Before he has even half opened it, she pushes through and begins rambling about mazes, portals and dragons. Isaac tells his friend to take a seat, and a deep breath, and start again. Rosa slows down and explains that she used eyes of Ender to track down an End portal, then dug down many layers to find the large stronghold that contains it.

PORTAL IS PRIMED

Rosa almost got lost in all those tunnels and dead-ends. She was happy to find the library, though she got slowed down looking at the books and fighting off skeletons. She is so tired she almost forgets to tell him the important part. She not only found the portal, but also activated it!

CAN THEY DEFEAT THE DRAGON?

It is clear that Rosa is excited. However, she also admits that she is quite nervous. The stronghold was dangerous enough; what about the Ender Dragon? She asks Isaac if he will help her to defeat the dragon and claim its egg as a trophy. Isaac agrees and the pair begin discussing everything they will need to complete their mission.

FRACTIONS (INCLUDING DECIMALS)

NOW FOR THE NETHER

Isaac will explore the **Nether** in search of new resources to help his and Rosa's quest to defeat the Ender Dragon. As a soul sand valley opens in front of him, he is pleased he came prepared. Before leaving home, Isaac enchanted his boots with **Soul Speed**; this means that he will not be slowed down as he sets out into this Nether biome.

SETTING FOOT ON SOUL SAND

A soul sand valley can only be found in the Nether. Hostile mobs loom large in this difficult landscape and, if that's not bad enough, the valley gets its name from the many soul sand blocks covering the ground. Soul sand slows walking speed – not helpful if being chased by a ghast.

STRIDE THROUGH THE LAVA

At the coasts, the soul sand gives way to gravel, which leads into oceans and lakes of lava. The only friendly mob, the striders, can be seen roaming through the ripples of molten rock, and they can be controlled if you have a saddle and warped fungus on a fishing rod. This makes for easier travel across the lava. Otherwise, mobs of skeletons and ghasts will keep you on your toes in the soul sand valley.

THE BLACK AND THE BEAUTY

There is beauty to be found. Often a traveller will stumble upon basalt pillars which tower out of the ground. Also breaking out from the soul sand are Nether fossils of long-gone creatures. Each of these can be harvested; the same goes for any blackstone found in bastions. The bastions are large blackstone castle structures. Piglin brutes that patrol the bastions don't like heroes trespassing and will attack if they spot them.

EQUIVALENT FRACTIONS

To simplify a fraction, divide the top and bottom of the fraction by the same number.

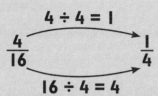

$$\frac{4}{16} \xrightarrow{4 \div 4 = 1} \frac{1}{4}$$
$$16 \div 4 = 4$$

To find an equivalent fraction, multiply the top and bottom of the fraction by the same number.

$$\frac{1}{4} \xrightarrow{1 \times 4 = 4} \frac{4}{16}$$
$$4 \times 4 = 16$$

As Isaac enters the soul sand valley, he can see down to a sea of lava. Walking carefully to investigate, he steps onto the gravel at the shore. Gravel can give way at any moment and Isaac does not want to end up in the red-hot stuff!

1

Complete these equivalent fractions.

$$\frac{2}{3} = \frac{\Box}{6} = \frac{6}{\Box} = \frac{\Box}{12} = \frac{10}{\Box} = \frac{\Box}{18}$$

2

Simplify each fraction to its lowest terms.

a) $\frac{21}{35} = $

b) $\frac{27}{63} = $

c) $\frac{18}{81} = $

d) $\frac{48}{64} = $

e) $\frac{24}{72} = $

f) $\frac{15}{45} = $

g) $\frac{25}{15} = $

h) $\frac{120}{1,000} = $

3

Write each set of fractions in order, starting with the smallest.

a) $\frac{4}{5}, \frac{2}{3}, \frac{7}{10}, \frac{11}{15}$ ....

b) $\frac{27}{40}, \frac{5}{8}, \frac{13}{20}, \frac{3}{5}$ ....

c) $5\frac{1}{3}, 5\frac{1}{5}, 5\frac{3}{10}, 5\frac{5}{6}$ ....

COLOUR IN HOW MANY EMERALDS YOU EARNED

FRACTIONS OF AMOUNTS

Ghasts fly overhead as Isaac walks back up the shore and to the safety of the soul sand. As he walks, he keeps his bow pulled tight in case a ghast notices him. He crouches now and again to pick mushrooms, which he enjoys in a stew.

1

For each shape, write the fraction that is shaded.

a)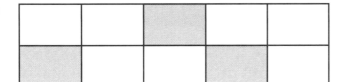

b)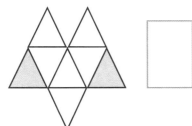

c)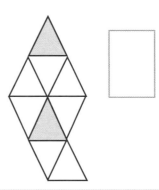

2

In this grid, a large rectangle is divided into smaller rectangles.

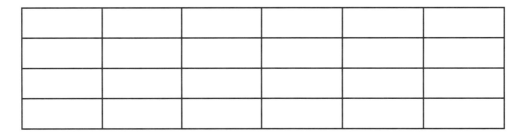

a) Colour: $\frac{1}{4}$ of the grid black $\frac{3}{8}$ of the grid green $\frac{1}{6}$ of the grid yellow

b) How many rectangles of each colour are there?

Black: ....   Green: ....   Yellow: ....

c) What fraction of the grid is not shaded?

Isaac has come across a bastion in the middle of a vast desert of soul sand. It stands from ground to the ceiling of the Nether. So many blocks tall! Isaac can't see a safe entry through its blackstone walls, so he begins to use soul sand and netherrack to build a column up to an opening. He can hear piglins inside but luckily he is wearing gold boots, which should help to avoid conflict.

3

a) Find $\frac{1}{4}$ of:

i) 16 soul sand ....

ii) 12 soul sand ....

iii) 20 soul sand ....

b) Find $\frac{1}{3}$ of:

i) 9 netherrack ....

ii) 24 netherrack ....

iii) 60 netherrack ....

Isaac approaches a piglin as he enters the bastion. He wants to barter with gold ingots for whatever the piglin can offer.

4

Isaac has 105 gold ingots. He gives $\frac{2}{5}$ of these gold ingots to the piglin.

How many gold ingots does he give to the piglin? .... gold ingots

---

5

The piglin drops two types of block: obsidian and crying obsidian. $\frac{1}{5}$ of the blocks are crying obsidian. 24 of the blocks are obsidian.

a) What fraction of the blocks are obsidian?

b) How many of the blocks are crying obsidian? .... blocks

c) How many blocks does the piglin drop in total? .... blocks

ADDING AND SUBTRACTING FRACTIONS

To add or subtract fractions, first write them as fractions with the same denominator.

$$\frac{5}{6} + \frac{1}{12} \qquad\qquad \frac{5}{6} = \frac{10}{12} \qquad\qquad \text{so} \qquad \frac{10}{12} + \frac{1}{12} = \frac{11}{12}$$

The bastion has blocks of gold placed in random areas. Isaac knows not to harvest the gold near piglins as they will get angry. He decides to climb the bastion and explore further in hope of finding some treasure chests.

 1

Complete these fraction calculations.

a)

$$\frac{1}{3} + \frac{1}{6} = \boxed{}$$

b)

$$\boxed{} + \boxed{} = \boxed{}$$

c)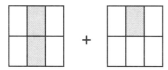

$$\boxed{} + \boxed{} = \boxed{}$$

d)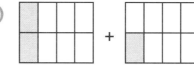

$$\boxed{} - \boxed{} = \boxed{}$$

Isaac has found a double chest in a corner of the bastion. As he approaches it to peek inside, a piglin brute swings its axe at him! Isaac stumbles backwards and begins defending himself. Other piglins notice the fight and help the brute by firing their crossbows.

2

Help Isaac fight the piglin brute. The brute has 50 health.
Each correct answer counts as 10 damage from Isaac's sword.

a) $\dfrac{1}{6} + \dfrac{1}{4} = \boxed{}$

b) $\dfrac{3}{4} + \dfrac{1}{3} = \boxed{}$

c) $\dfrac{4}{7} + \dfrac{1}{3} = \boxed{}$

d) $\dfrac{7}{15} - \dfrac{1}{5} = \boxed{}$

e) $\dfrac{9}{12} - \dfrac{1}{4} = \boxed{}$

f) $\dfrac{9}{10} - \dfrac{3}{4} = \boxed{}$

3

Three piglins fire arrows at Isaac.

The first piglin fires $\dfrac{1}{5}$ of the arrows. The second piglin fires $\dfrac{1}{3}$ of the arrows.

What fraction of the arrows does the third piglin fire?

$\boxed{}$

4

Work out:

a) $1\dfrac{5}{8} + 3\dfrac{1}{4} = $ ....

b) $3\dfrac{2}{3} + 1\dfrac{1}{3} = $ ....

c) $2\dfrac{1}{12} - 1\dfrac{2}{3} = $ ....

d) $2\dfrac{5}{6} - \dfrac{4}{5} = $ ....

5

 Find the missing numbers in these calculations.

a) $3\dfrac{2}{3} - 1\dfrac{\boxed{}}{5} = 1\dfrac{13}{15}$

b) $\dfrac{4}{5} + 1\dfrac{9}{\boxed{}} = 2\dfrac{7}{10}$

c) $\dfrac{11}{12} - \dfrac{\boxed{}}{5} = \dfrac{19}{60}$

COLOUR IN HOW MANY EMERALDS YOU EARNED

MULTIPLYING AND DIVIDING FRACTIONS

To multiply fractions, multiply the numerators and multiply the denominators:

$$2 \times 1 = 2$$

$$\frac{2}{3} \times \frac{1}{4} = \frac{2}{12} = \frac{1}{6}$$

$$3 \times 4 = 12$$

To divide a fraction by a whole number, write the whole number as a fraction, flip it and replace ÷ with ×

$$2 \times 1 = 2$$

$$\frac{2}{3} \div 4 = \frac{2}{3} \div \frac{4}{1} = \frac{2}{3} \times \frac{1}{4} = \frac{2}{12} = \frac{1}{6}$$

$$3 \times 4 = 12$$

Flip the fraction and replace ÷ with ×

Isaac fights off the piglins and gets back to looting the chests he has found. There are lots of items inside. He must decide which to take and which to leave, as he can't carry it all. Isaac still has more of the valley to explore and will need to save some space.

1

Work out:

a) $\frac{1}{8} \times \frac{1}{5} =$

b) $\frac{1}{25} \times \frac{1}{4} =$

c) $\frac{2}{3} \times \frac{1}{6} =$

d) $\frac{3}{4} \times \frac{3}{5} =$

2

Work out:

a) $\frac{1}{2} \div 2 =$

b) $\frac{1}{3} \div 4 =$

c) $\frac{1}{5} \div 2 =$

d) $\frac{1}{6} \div 5 =$

e) $\frac{3}{8} \div 4 =$

f) $\frac{5}{6} \div 5 =$

Isaac has separated the items he found inside the chests. There were lots of iron and gold nuggets, along with a golden apple, which will help in tough fights, and a music disc called 'Pigstep'. Isaac enjoys listening to music and must build a jukebox when he gets home.

3

Find the missing numbers in these fraction calculations.

a) $\dfrac{1}{4} \times \dfrac{\square}{\square} = \dfrac{1}{16}$

b) $\dfrac{\square}{\square} \times \dfrac{1}{4} = \dfrac{1}{64}$

c) $\dfrac{2}{5} \times \dfrac{\square}{\square} = \dfrac{2}{15}$

d) $\dfrac{1}{6} \div \square = \dfrac{1}{30}$

e) $\dfrac{1}{\square} \times \dfrac{1}{3} = \dfrac{1}{6} \times \dfrac{\square}{2}$

f) $\dfrac{5}{\square} \div 2 = \dfrac{1}{4} \times \dfrac{\square}{2}$

4

Isaac has already used $\dfrac{1}{5}$ of his food.

He uses the rest of his food to prepare three identical snacks.

What fraction of his food does he use for each snack?

5

In a chest full of potions:

- $\dfrac{3}{4}$ of the potions are Potions of Healing

- half of the Potions of Healing are level II.

What fraction of the potions in the chest are Potions of Healing level II?

COLOUR IN HOW MANY EMERALDS YOU EARNED

DECIMALS

To convert a fraction to a decimal, divide the numerator by the denominator:

$$\frac{4}{5} = 4 \div 5 = 5\overline{)\begin{array}{c} 0\ .\ 8 \\ 4\ .\ 0 \end{array}}$$

To convert a decimal to a fraction, use place value:

$0.03 = \dfrac{3}{100}$ $0.55 = \dfrac{55}{100} = \dfrac{11}{20}$

Isaac is delighted with the number of iron and gold nuggets he has found because they can be used to craft different things. Help him to carry out some fraction and decimal conversions to work out how he will use them.

1

Write each fraction as a decimal.

a) $\dfrac{1}{2} =$ ....

b) $\dfrac{3}{4} =$ ....

c) $\dfrac{1}{5} =$ ....

d) $\dfrac{1}{10} =$ ....

e) $\dfrac{5}{8} =$ ....

f) $\dfrac{7}{20} =$ ....

2

Write each decimal as a fraction in its lowest terms.

a)   0.09 =

b)   0.71 =

c)   0.437 =

d)   0.103 =

e)   0.24 =

f)   0.6 =

g)   0.35 =

h)   0.375 =

i)   0.45 =

Moving through the Nether is hard work. Isaac must keep stopping to fight, then pause again to eat. Life in this dimension would be much easier with some minecarts and tracks!

3

Write the decimal values in the boxes on each number line.

a)

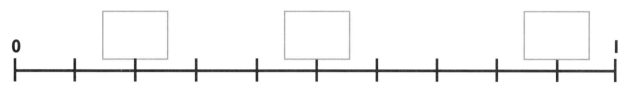

0 ___ ___ ___ 1

b)

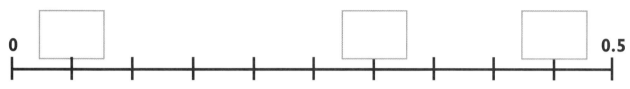

0 ___ ___ ___ 0.5

c)

0 ___ ___ ___ 0.05

4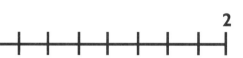

Draw labelled arrows to mark these decimals on the number line.

A = 1.55   **B** = 0.05   **C** = 0.70

0 _____ 2

5

Write these fractions and decimals in order, starting with the smallest.

0.76 $\frac{1}{3}$ 0.3 $\frac{2}{5}$ 0.04

PERCENTAGES

$5\% = \dfrac{5}{100} =$ ⟨ $\dfrac{1}{20}$ as a simplified fraction
0.05 as a decimal

$10\% = \dfrac{10}{100} =$ ⟨ $\dfrac{1}{10}$ as a simplified fraction
0.1 as a decimal

To find 10% of an amount, divide by 10.

Although he has found some terrific items and materials in the soul sand valley, Isaac has used up some of the equipment and food supplies he set out with. He has less than half of some of his resources left.

1

Work out:

a)   10% of 200 arrows   .... arrows

b)   25% of 80 melon slices   .... melon slices

c)   50% of 150 torches   .... torches

2

Write the fraction and decimal equivalents for each percentage.

a)   90% = [   ] = ...

b)   35% = [   ] = ...

c)   75% = [   ] = ...

d)   40% = [   ] = ...

As Isaac reaches the top of a hill made of soul sand, he looks down onto a large fossil sticking out of the ground. It is huge! Whatever this monster was, Isaac is glad he doesn't have to fight it. He approaches the bones and uses his pickaxe to see if they can be harvested.

3

The price of a set of bones is 160 emeralds.

The price rises by 5%

By how many emeralds does the price increase? .... emeralds

4

Isaac and Rosa did a quiz about the Nether.

Isaac scored 60% in the quiz.

Rosa scored $\frac{13}{20}$ in the quiz.

Who got the better score in the quiz? ....

5

 In pounds and pence, these are the prices of some expedition equipment:

DIAMOND BOOTS	DIAMOND LEGGINGS	DIAMOND HELMET

£24	£51	£27.50

The prices are reduced by 20%. Work out the new price of each item.

Diamond boots: £.....................

Diamond leggings: £.....................

Diamond helmet: £.....................

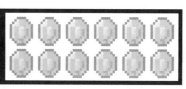
COLOUR IN HOW MANY EMERALDS YOU EARNED

MIXED FRACTION, DECIMAL AND PERCENTAGE PROBLEMS

As much as Isaac would love to keep pushing forward and find more amazing items, he must return home. His inventory is full to bursting and his weapons are wearing down. The last thing to do before leaving is to collect some soul sand.

1

Shade $\frac{3}{4}$ of this shape.

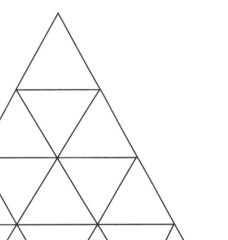

2

Write these values in order, from smallest to largest.

$\frac{8}{10}$ 0.08 $\frac{3}{4}$ 60%

Isaac's journey back to the Nether portal is dangerous. Endermen are flitting back and forth around him. He must avoid looking directly at the Endermen, otherwise they will attack. Isaac has a trick up his sleeve, though. Placing a carved pumpkin over his head enables him to avoid eye contact with them.

3

Isaac fought 400 mobs last year.

- 10% of the mobs were creepers
- 60% of the mobs were zombies
- 25% of the mobs were skeletons
- 5% of the mobs were phantoms

a) How many of the mobs Isaac fought were:

creepers? .... skeletons? ....

zombies? .... phantoms? ....

b) How many more skeleton mobs than creeper mobs did Isaac fight? ....

4

Write these fractions in order, starting with the largest.

$\frac{2}{3}$ $\qquad$ $\frac{4}{5}$ $\qquad$ $\frac{7}{10}$ $\qquad$ $\frac{9}{15}$

5

Isaac has some potatoes left over.

$\frac{1}{4}$ of the potatoes are baked.

15 potatoes are not baked.

a) What fraction of the potatoes are not baked?

b) How many potatoes are baked? .... potatoes

c) In total, how many potatoes are left over? .... potatoes

Isaac reaches the portal without any more skirmishes. As he steps through and the purple shimmer of the gate surrounds him, he feels very satisfied with his expedition. He explored a lot of the soul sand valley and collected many items. Isaac steps out, tired from his exploits, and is relieved to see his house. Answer these questions while he settles back in at home.

6

Fill in the missing numbers in these fraction calculations.

a) $\dfrac{1}{3} \times \dfrac{1}{2} = \dfrac{\Box}{\Box}$

b) $\dfrac{1}{4} \times \dfrac{\Box}{\Box} = \dfrac{1}{8}$

c) $\dfrac{\Box}{\Box} \times \dfrac{1}{3} = \dfrac{1}{12}$

d) $\dfrac{1}{5} \times \dfrac{\Box}{\Box} = \dfrac{1}{20}$

7

The value in each cube is the total of the two cubes below it. Fill in the blank cubes.

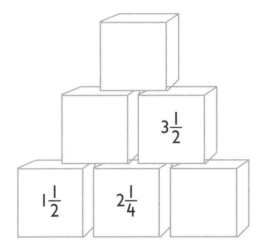

8

Isaac has a lucky number. 30% of this number is 18.

a) What is Isaac's lucky number? ....

b) What is 65% of Isaac's lucky number? ....

ADVENTURE ROUND-UP

BRUTAL BATTLES

Isaac had a lot of fun in the soul sand valley even though he was in danger many times. First it was the perilous sea of lava then, when he found the bastion, a group of piglins attacked him. He had to fight using both his sword and his bow. Next time he will take a shield. He did get to loot the treasure from the bastion, including an interesting music disc. Isaac is relieved that he packed the carved pumpkin, which helped him to return safely to the Nether portal.

A PIECE OF HISTORY

Finding the fossil was an exciting moment; such big bones towering out of the sand! Isaac wonders what kind of animal they might have belonged to. Maybe a dragon? Now he is home and ready for food and a rest, but Rosa knows where the stronghold is and, together, they want to make a plan of how to take down the Ender Dragon. First though, he tucks into a stack of cookies.

RATIO, PROPORTION AND ALGEBRA

THE END IS IN SIGHT

Having travelled back to the stronghold, Rosa and Isaac look down on the End portal. After placing the eyes of Ender, the portal comes to life with a space filled with distant stars. This is a one-way trip, for now. The only way to exit is to defeat the Ender Dragon. Both of our heroes are ready and prepared. They nod and step forward, falling into endless space.

THE EXTRAORDINARY END

The End is a strange dimension. Like the Nether, it is filled with danger. Only two mobs live on the large, central island: Endermen and the Ender Dragon. Endermen walk and teleport all over the ground in sparks of purple. In the sky above, the Ender Dragon is quite a sight with its massive wings and long body.

A FLOATING ISLAND

The central island of the End floats in dimensional space. Falling off the edge into the Void has fatal consequences. The ground is made from End stone, a mix of yellow and green. This material is easy to break and harvest. Rising in a circle towards the centre are 10 obsidian pillars each topped with an End crystal, two of which are caged. The End crystals can heal the Ender Dragon.

THE ULTIMATE TEST

At the very centre of the island and the pillars is the exit portal. This is the ticket back to normality, but the portal remains inactive until the Ender Dragon is dead.

RATIO

The ratio of caged End crystals to uncaged End crystals is 1 : 4

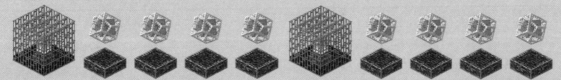

This means that for every 1 caged End crystal, there are 4 uncaged End crystals.

You can simplify a ratio by dividing both values by the same number.

2 : 8 can be simplified
to 1 : 4 like this:

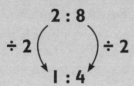

Rosa and Isaac plunge through the End portal and find themselves on the central island. Above them, they can hear the Ender Dragon soaring through the air and roaring fiery breath. Time for a quick check on equipment.

1

Rosa has pink splash Potions of Healing. Isaac has purple splash Potions of Slow Falling.

They put the potions in a row, in the ratio of   **1 purple : 3 pink**

Colour the splash potion bottles in this ratio.

2

Simplify these ratios.

a)   4 : 12 ....   b) 8 : 16 ....   c) 3 : 24 ....   d) 5 : 25 ....

3

 Rosa is sorting her arrows into a pattern.

The pattern is 3 grey Arrows of Weakness followed by 5 green Arrows of Poison.

Rosa uses 45 green Arrows of Poison in total.

How many grey Arrows of Weakness does she use? .... Arrows of Weakness

**COLOUR IN HOW MANY
EMERALDS YOU EARNED**

SCALING

These two triangles are *similar*. This means they are the same shape but have different measurements.

All the side lengths of triangle **B** are twice those of triangle **A**.

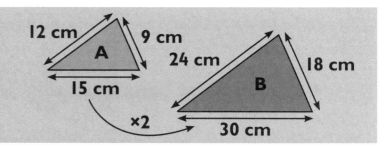

Rosa plans to fire arrows from the ground up to the End crystals. Meanwhile, Isaac will build up to the crystals using cobblestone and smash them with his netherite sword. This will stop the Ender Dragon from being healed during the fight.

1

Isaac notices he is low on energy. He has some beetroot to eat.

6 beetroot replenishes 7.2 energy.

How much energy do these numbers of beetroot replenish?

a)   3 beetroot   .... energy

b)   12 beetroot   .... energy

2

Rosa test fires an arrow. It lands 6.5 m away.

Then she fires a second arrow. It goes 4 times as far.

How far away does the second arrow land?   .... m

3

Rosa test fires more arrows at three targets. The targets are three similar rectangles:

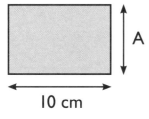

A

10 cm

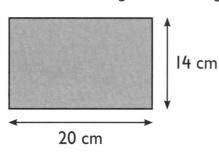

14 cm

20 cm

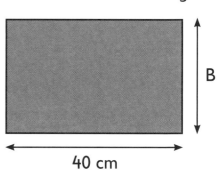

B

40 cm

a)   What is the length of side A?   .... cm

b)   What is the length of side B?   .... cm

Isaac and Rosa will need to deal with the Endermen before focussing on the Ender Dragon. This will help keep them safe, plus they can use any Ender pearls to teleport away from danger when the big battle begins.

4

Here are two sizes of cuboid:

A big cuboid measures
49 cm × 21 cm × 14 cm

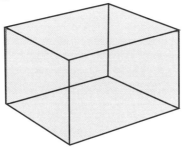

A small cuboid measures
7 cm × 3 cm × 2 cm

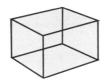

How many times longer are the lengths of a big
cuboid than those of a small cuboid?   .... times longer

5

 Isaac and Rosa stocked up on apples and carrots for this quest.
These were the prices they paid in pounds and pence:

Apples	Carrots
£1.60 per kg	70p per kg

a)   Isaac bought 1.5 kg of apples.

How much did he spend?   £ ....

b)   Rosa spent £2.10 on carrots.

How many kilograms of carrots did she buy?   .... kg

UNEQUAL SHARING

Rosa and Isaac share 45 pieces of loot in the ratio 1 : 2

We can show this as a picture: Rosa's share: Isaac's share:

Altogether, there are 3 shares.

Each share is 45 ÷ 3 = 15

Rosa: 1 × 15 = 15 pieces   Isaac: 2 × 15 = 30 pieces

Rosa and Isaac defeat lots of Endermen. They each fight using their bows and swords. With fewer Endermen around, they can focus on the Ender Dragon.

1

The defeated Endermen have dropped lots of Ender pearls.

Isaac and Rosa collect the Ender pearls in the ratio 1 : 3

Complete this table to show how many Ender pearls they each collect.

Isaac	Rosa
1	3
2	
5	
10	

2

Isaac and Rosa share 50 Ender pearls in the ratio 4 : 1

How many Ender pearls are in each share?

Isaac: ....

Rosa: ....

A roar rumbles through the sky! The Ender Dragon has noticed the heroes and swoops down to get a closer look. Rosa and Isaac run for cover in order to prepare for an epic battle.

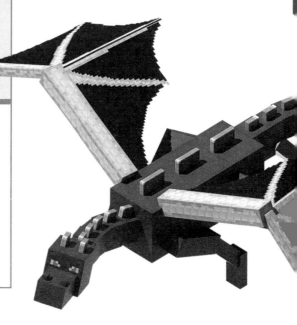

3

Before setting off on this quest, Rosa and Isaac shared out 160 emeralds.

Rosa got three times as many as Isaac.

How many emeralds did Rosa get?

.... emeralds

4

 Rosa and Isaac spend 120 seconds making their final battle preparations.

During those 120 seconds:

- They spend three times longer discussing their strategy than they do checking their weapons.
- They spend twice as long eating as they do checking their weapons.

How many seconds do they spend on each part of their preparation?

Discussing strategy: .... seconds

Eating: .... seconds

Checking weapons: .... seconds

5

 Isaac quickly gobbles a cake made of wheat, sugar and eggs in the ratio 4 : 6 : 5

These ingredients weight 900 g in total.

How much sugar is in the cake? ....g

COLOUR IN HOW MANY EMERALDS YOU EARNED

57

FORMULAE AND EQUATIONS

Rosa has a secret number. She adds 12 to her number. The answer is 20.

You can write this using algebra. Use *n* for the secret number: $n + 12 = 20$

You can solve this equation to find that $n = 8$.

Rosa hides behind an obsidian pillar. Her plan is to keep stepping out to fire arrows up at the End crystals. She begins firing arrow after arrow, with the first few falling short of the target. Soon she finds her range and starts to take out the crystals, one by one.

1

Work out the value of the letter in each equation.

a) $x + 3 = 9$ $x = $ ....

b) $y - 5 = 18$ $y = $ ....

c) $2n = 14$ $n = $ ....

2

These pictures represent numbers:

 +  = 30

 –  = 28

Find the number represented by each picture.

 = ...  = ...

Meanwhile, Isaac starts building tall columns of cobblestone in order to reach and break the iron bars surrounding the caged End crystals. Rosa will then fire arrows at the crystals. Isaac is glad he has Potions of Slow Falling and boots enchanted with Feather Falling. This means that when he has broken the bars, he can jump back down to the ground safely.

3

Isaac uses this formula to work out how much time he will need for building the cobblestone columns:

$$T = 15n$$

T is the time needed in seconds and n is the number of obsidian pillars.

Work out how much time he will need for 2 obsidian pillars.

.... seconds

4

Work out the value of $4x - 3$ when $x = 6$.

$4x - 3 =$ ....

5

It takes Rosa 6 attempts to hit the last End crystal.

a and b are positive whole numbers.

$$a - b = 6$$

Give three possible pairs of values for a and b.

$a =$ .... and $b =$ ....

$a =$ .... and $b =$ ....

$a =$ .... and $b =$ ....

LINEAR NUMBER SEQUENCES

A linear number sequence goes up (or down) in equal sized steps.

With all the End crystals broken, the heroes can concentrate on the Ender Dragon. The dragon will no longer heal, so they can steadily attack until they secure victory.

1

What are the next three terms in each sequence?

a) 24   15   6   ....

b) 23   31.5   40   ....

2

The rule for this sequence is 'add 8':   3   11   19   27

a) This sequence shows the number of successful arrow strikes on the Ender Dragon. Write the rule for the sequence.

 7   11   15   19   23

....

b) This sequence shows how the Ender Dragon's health changes. Write the rule for the sequence.

 100   92   84   76

....

The dragon is weak! Rosa and Isaac rush in with their swords to finish the fight.

3

a) Write the missing numbers in this sequence of sword strikes.

....   36   45   63

b) The dragon is defeated! It rises into the air and explodes in a shower of experience orbs. Write the missing numbers in this sequence of experience orbs.

56   84

COLOUR IN HOW MANY EMERALDS YOU EARNED ⬡⬡⬡⬡⬡⬡

ADVENTURE ROUND-UP

DRAGON DEFEATED

The End portal is now ready and activated. Treading carefully, Rosa and Isaac take steady breaths and walk towards it. On top of a small pillar above the portal is a dragon's egg. It takes a little time and skill, but the pair manage to remove the egg to take home as a trophy of their battle with the Ender Dragon.

HEROES HEAD HOME

Peace has returned and all is quiet. The only ones left standing are the heroes of the moment. Both manage a smile and straighten up to their full height, bursting with pride. Stepping through the portal again, everything goes dark.

IT WASN'T A DREAM!

Rosa emerges from the portal and travels home, before going to sleep. When she wakes up, for a moment she wonders if the battle was a dream. But no, the dragon's egg is right there. With the Ender Dragon beaten, Rosa and Isaac achieved their goal.

MEASUREMENT

ISAAC INVESTIGATES

Isaac, fully rested after defeating the Ender Dragon, is feeling more confident than ever. He has now decided to investigate another biome in the Nether – the warped forest. However, when he arrives, he realises he didn't need to bring weapons...only his crafting tools and his enthusiasm for learning.

PEACE IN THE NETHER?

The warped forest is a gloomy and mysterious biome found in the Nether, but is unusually peaceful for that dimension. Endermen will pass by without causing any trouble unless an explorer angers them by making eye contact. At the coast, far out in the lava, the striders walk with their babies.

PLANTS APLENTY

From the cyan coloured trees to the twisting vines and massive fungi, there is plenty to harvest or admire. While it may be dark all around the edges of the forest, the biome is lit naturally by shroomlights, which grow among the leaves of the trees. However, it is still easy to get lost.

WOOD OF THE WARPED FOREST

Bastions and fortresses sometimes tower over the landscape and are potential treasure troves. The warped forest also offers budding builders an opportunity to harvest a blue-green colour of wood.

PERIMETER AND AREA

Perimeter is the distance around the outside of a 2-D shape.

Area is the space inside a 2-D shape.

Isaac begins his exploration of the warped forest by building straight up on a column of cobblestone. This gives him a great view of his surroundings.

1

The zone of warped forest that Isaac is investigating is a rectangle 3 km long and 2 km wide.

Calculate the perimeter and area of the zone.

Perimeter = ... km

Area = ... km²

2

This is a plan of a smaller part of the warped forest:

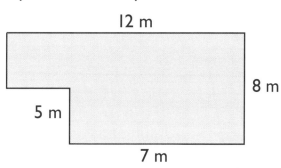

12 m

8 m

5 m

7 m

a) Work out the length of fencing needed to go all around this part of the warped forest. .... m

b) Work out the area of this part of the warped forest. .... m²

3

Isaac has 20 lengths of stone wall. Each length is 2 metres long.

He uses them to make a four-sided enclosure with the largest possible area.

What is the largest possible area he can make? .... m²

COLOUR IN HOW MANY EMERALDS YOU EARNED

AREA OF PARALLELOGRAMS AND TRIANGLES

Area of a parallelogram is:

base × height   or $b × h$

Area of a triangle is:

$\dfrac{\text{base × height}}{2}$ or $\dfrac{b × h}{2}$

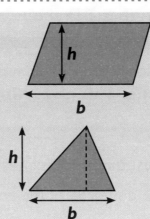

From his lookout position, Isaac spots some places for closer investigation. He makes some rough sketches. Find the area of the shapes he has drawn.

1

Work out the area of this parallelogram:

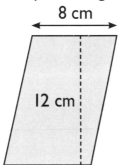

8 cm

12 cm

.... cm²

2

Work out the area of each triangle.

a)

5 cm

12 cm

.... cm²

b)

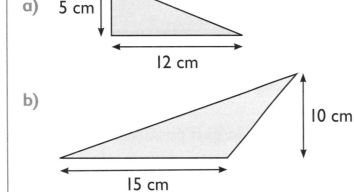

10 cm

15 cm

.... cm²

Isaac spots an Enderman roaming between the trees.

3

Here is a sketch of an Enderman using 2-D shapes.

What is the area of this sketch of the Enderman?

.... cm²

5 cm

5 cm

4 cm

4 cm

4 cm

4 cm

17 cm

13 cm

1 cm

1 cm

4

 Find the length of the base of a parallelogram that has these measurements:
area 600 cm² and height 20 cm

base = ... cm

COLOUR IN HOW MANY EMERALDS YOU EARNED

VOLUME

Volume is the amount of space inside a 3-D shape.

Volume of a cuboid = length × width × height or *l* × *w* × *h*

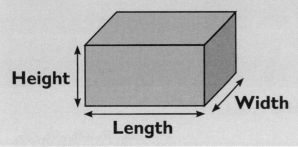

Height

Width

Length

Although Isaac can't sleep in the Nether, he has decided to build a small hut where he will store a chest. This can hold precious items in case anything dangerous happens while he is exploring.

 1

Here are two possible styles of cuboid-shaped hut that Isaac could build. Work out the volume of each one.

a)

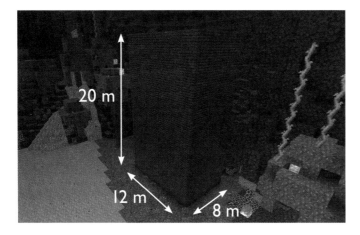

20 m

12 m

8 m

.... m³

b)

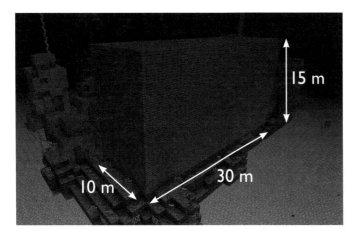

15 m

30 m

10 m

.... m³

Isaac is going to take a closer look at the striders which are found at a lava lake. He uses cobblestone to build a small platform out onto the lake. He has a saddle ready, hoping he will be able to ride a strider through the lava.

2

Here is Isaac's platform made from 1 m blocks. The platform is not finished.

When the platform is complete, it will be a cuboid 8 m long, 4 m wide and 3 m high.

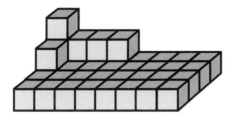

How many more blocks does Isaac need to finish the platform? ....

3

What is the volume of a cube with sides of 6 cm?

.... cm³

4

A cube has volume 8 cm³.

What is the length of one side of the cube?

.... cm

5

 Rosa builds a cuboid using centimetre cubes. The cuboid is 8 cm long, 5 cm wide and 6 cm high.

She can also use the same number of cubes to make a second cuboid.

What could the length, width and height of Rosa's second cuboid be?

Length: .... cm

Width: .... cm

Height: .... cm

CONVERTING UNITS OF MEASUREMENT

Remember:	1 km = 1,000 m	1 inch is approximately 2.5 cm
	1 m = 100 cm	1 mile is approximately 1.6 km
	1 cm = 10 mm	1 pint is approximately half a litre
	1 l = 1,000 ml	
	1 kg = 1,000 g	

Isaac stands on the edge of his platform and waits. After some time, a strider gets close enough for him to fasten the saddle to it. He uses a warped fungus on a fishing rod to guide the creature, and they gallop off across the surface of the lava! While Isaac enjoys the ride, answer these questions.

1

In each pair of measurements, circle the larger one.

a)   200 millilitres   1 pint

b)   12 inches   35 centimetres

c)   1 mile   1 kilometre

2

Convert these measurements to the unit given.

a)   6.5 kg = ... g

b)   7.9 cm = ... mm

c)   9.35 m = ... cm

d)   3,975 m = ... km

e)   10,825 ml = ... l

f)   8,025 g = ... kg

3

💜 Isaac would ride the strider 300 kilometres if he could!

Approximately how far is this distance in miles?   .... miles

COLOUR IN HOW MANY EMERALDS YOU EARNED

ADVENTURE ROUND-UP

A SUPER STRIDER RIDE

Isaac's mini-adventure comes to an end as the strider guides him back to his cobblestone platform and the safety of land. He removes the saddle, feeds the strider some warped fungus and says goodbye. As he watches the strider wander off, he marvels at the wonders of the Nether. For a place so often filled with danger, there are still places to stop and admire.

SHROOMLIGHTS SECURED

Isaac begins packing up his chest and items. In the warped forest he collected lots of wood and fungi. He also harvested lots of shroomlight blocks. They will make a lovely addition to his garden, lighting up the surroundings. Smiling with success, Isaac steps through the portal and returns home to tell Rosa all about it.

GEOMETRY

RUMOURS LEAD TO A RETURN

Rosa has heard rumours that cities can be found in the End. Treasures and unusual creatures are said to lay undisturbed. Without delay, Rosa arms herself with equipment, potions and food. It's back to the End!

NOT QUITE THE END

The End is the lair of the Ender Dragon. But what happens when the Ender Dragon is defeated, and the heroes win the battle? Beyond the dragon's island, with its obsidian pillars, are the outer islands. These End islands are a treasure trove for any explorer.

SKYSCRAPER CITIES

Throughout the islands there are sights to see, including End cities and End ships. These structures are made from a material called purpur, perhaps after its lilac colouring. The cities reach upwards rather than outwards and the End ships are moored to their upper floors.

UP FOR THE CHALLENGE?

The End islands are the only place where you will find elytra – a set of wings which will allow Rosa or Isaac to soar through the sky like a bird. A hostile mob called shulkers are native to this dimension. They hide inside shells and guard the cities, ensuring only the bravest explorers find glory.

2-D SHAPES

In a circle:

- the *circumference* is the distance around the outside edge
- the *radius* is the distance from the centre to the outside edge
- the *diameter* is the distance from edge to edge, through the centre.

Many of the islands of the End are separated by large distances. Rosa knows that she will need to use Ender pearls in order to move between them.

 1

Tick the circle that has a dashed line showing the diameter of the circle.

☐   ☐   ☐   ☐

2

What is the radius of this circle?

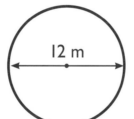

12 m

.... m

Rosa exits the same End portal which brought her face to face with the Ender Dragon. Ahead of her, she sees a new type of gateway, made from bedrock. It is shaped like an octahedron and, at the very centre, is a small opening. A magenta beam lights the sky above the gateway. Rosa pulls back her arm and throws an Ender pearl right into the middle of the gateway.

3

As Rosa is teleported, all kinds of 2-D shapes flash before her eyes.

   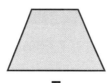   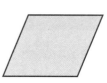

  A **B** **C** **D**

Write the letters of the correct shapes beside each statement.

a)   This shape has exactly two pairs of parallel sides. ....

b)   These shapes have no right angles. ....

c)   These shapes are quadrilaterals. ....

d)   This shape has exactly one pair of parallel sides. ....

4

Two sides of a square are drawn on this grid.

Complete the other two sides of the square.

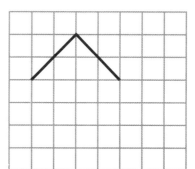

5

These red, blue and yellow circles have the measurements shown.

Diameter   Radius   Radius
2 cm   0.5 cm   0.75 cm

Rosa draws 7 red circles, 3 blue circles and 4 yellow circles edge to edge in a straight line.

What is the length of her line of circles? .... cm

Landing with a thump, Rosa emerges from the gateway. Looking around her, she is standing in front of an End city and an End ship! Laughing with joy, Rosa jumps into the air and skips off to the city, not knowing that danger awaits her. As she begins to enter the city, she passes a strange box. Suddenly, the box opens and from inside a small creature – a shulker – fires a projectile at her. Rosa tries to dodge it, but it follows her.

6

Four circles fit exactly into a rectangle.

Each circle has radius 2.5 cm.

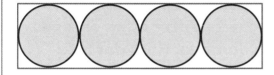

Find the perimeter of the rectangle. .... cm

7

On this square grid, draw a pentagon with three right angles.

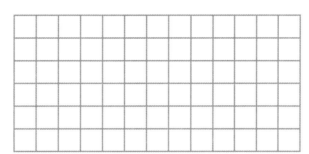

8

 These five identical circles are inside a hexagon, so that the sides of each just touch.

The perimeter of this hexagon is 120 cm.

Work out the radius of one of the circles.

.... cm

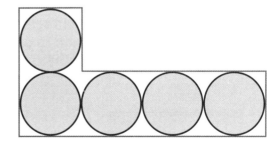

3-D SHAPES AND NETS

A net is a 2-D shape that folds up to make a 3-D shape.

A 3-D shape has faces, edges and vertices.

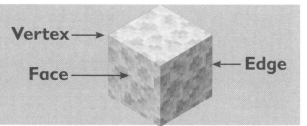

Vertex ⟶

Face ⟶

⟵ Edge

The shulker's projectile hits Rosa as she tries to run. Taking damage, she is shocked to find herself floating up into the air. As Rosa rises, she sees built structures that from above look like 3-D shapes and nets. After a few seconds, she comes crashing to the ground, taking yet more damage. Rosa darts to safety to heal.

1

Here are three 3-D shapes:

Triangular prism	Cuboid	Square-based pyramid

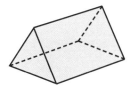 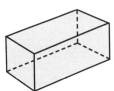 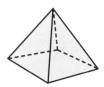

a)   Which two of these shapes have five faces?

_____ and _____

b)   Which two of these shapes have parallel faces?

_____ and _____

2

What 3-D shape does each net make?

a)

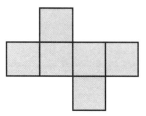

b)

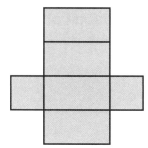

_____   _____

Rosa's health has almost regenerated. She wonders what kind of projectiles she will encounter next. Could they be different 3-D shapes?

3

Which of these shapes is the net of a triangular prism?

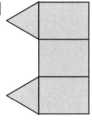 A 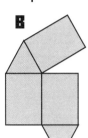 B C D E

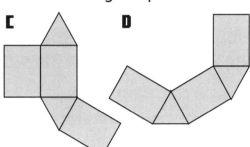 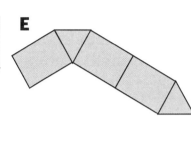

Shape ....

4

This is the net of a 3-D shape.

How many faces, edges and vertices does the 3-D shape have?

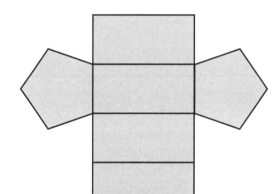

Faces: ....

Edges: ....

Vertices: ....

5

 Rosa describes a 3-D shape:

'It has 6 faces, 12 edges and 8 vertices.

Its faces are two different 2-D shapes.'

What shape has Rosa described? ....

MISSING ANGLES

The angles in a triangle add up to 180°.

The angles in a quadrilateral add up to 360°.

A regular polygon has all sides equal and all angles equal.

Rosa steps out of her hiding place. She can see a shulker up ahead. Its shell opens as it peeks out, before closing again. What if she can shoot an arrow into the shell opening? She would have to get the angle exactly right, but she has had plenty of practice.

1

Rosa needs to work out her angles of attack.

Calculate the size of angle x in each diagram.

a)
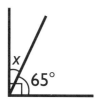

Angle x = ...°

b)

Angle x = ...°

c)
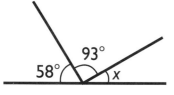

Angle x = ...°

d)
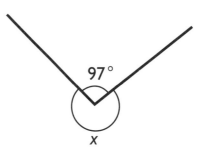

Angle x = ...°

2

Work out the size of angle x.

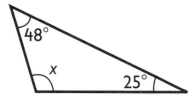

Angle x = ...°

Managing to escape the remaining shulkers, Rosa ducks inside a tall tower. She stops in her tracks. There are hardly any stairs. Instead, purpur blocks and sticks poke out from the walls. After many jumps and a few stumbles, Rosa makes it to the top and looks out onto the End islands.

3

Find the size of angle y.

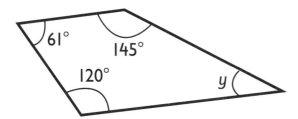

Angle y = ...°

4

Here is a regular polygon, sitting on a straight line.

Work out the size of angle x.

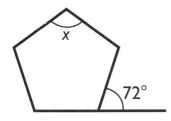

Angle x = ...°

5

Work out angles x and y in each diagram.

a)

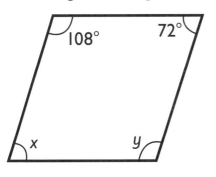

Angle x = ...°

Angle y = ...°

b)

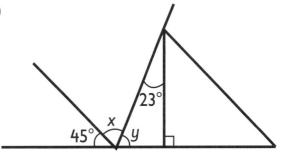

Angle x = ...°

Angle y = ...°

CO-ORDINATES

> From where Rosa stands at the top of the tower, she can see far into the distance. Opening her notebook, she sketches a rough layout of the area to make sure she can eventually find her way back.

1

Rosa can see an Enderman, the End ship and the End gateway.

The three dots on this co-ordinate grid show their locations.

They are three vertices of a square.

a) Write down the co-ordinates of these vertices.

(.................... , ...)

(.................... , ...)

(.................... , ...)

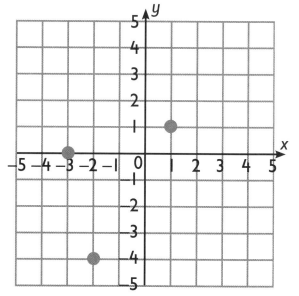

b) What are the co-ordinates of the fourth vertex of this square?

(.................... , ...)

2

Rosa can see three more landmarks, at these co-ordinates:

(−2, −3)   (3, −1)   (0, 1)

They form three vertices of a parallelogram.

a) Mark these points on the grid.

b) Write down the co-ordinates of the fourth vertex of the parallelogram.

(.................... , ...)

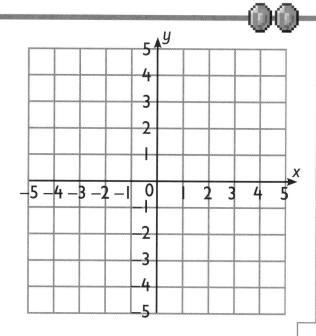

What a strange place this is! Rosa climbs carefully back down the tower to investigate the End ship. The ship seems to be facing away from the islands, as if it were about to leave on a long voyage. Rosa spots several shulkers on the ship. With bravery, she approaches with her sword ready.

3

This diagram shows the position of Rosa and a shulker.

What are the co-ordinates of the point halfway between Rosa and the shulker?

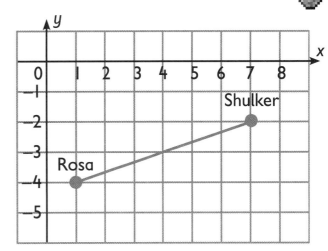

(...................., ...)

4

a) Rosa marked these points on her sketch of the surroundings:

A (−5, 1)   B (−3, 5)   C (0, 5)   D (3, 1)

Plot the points on the grid.

b) Join the points with straight lines: A to B, B to C, C to D and D to A.

What shape is ABCD?

....

c) Find the co-ordinates of the midpoint of line AC.

(...................., ...)

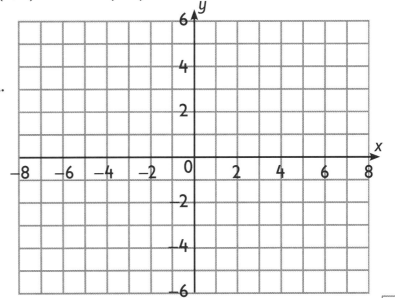

5

 A straight line joins the points (−2, −6) and (4, 2).

Find the co-ordinates of the midpoint of this line.   (...................., ...)

COLOUR IN HOW MANY EMERALDS YOU EARNED

TRANSLATION AND REFLECTION

A translation slides a shape across the co-ordinate grid:

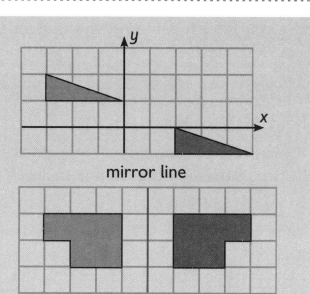

mirror line

A reflection flips a shape in a mirror line:

There are three shulkers on the End ship in total – one on the deck, one at the stern and one guarding what looks like a treasure room. Using all her skill, Rosa picks off each shulker with her arrows, allowing her to explore.

1

Describe the translation that takes shape 1 to shape 2.

.... squares right and

.... squares ....

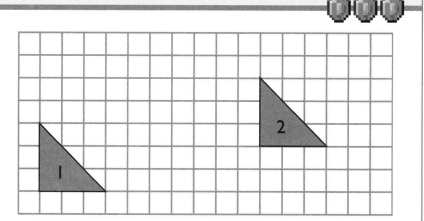

2

Reflect this triangle in the *x*-axis.

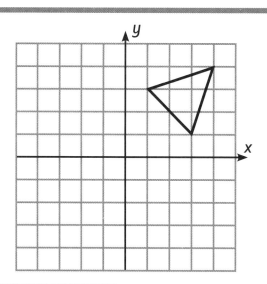

The ship is magnificent! Rosa vows to try to build something so grand when she returns home. Stepping into the treasure room, she finds two chests, which lie close to where the shulker stood guard. Opening them, Rosa finds diamonds, enchanted tools and loads of gold ingots!

3

The grid shows five shapes labelled A, B, C, D and E.

a) Which two rectangles show a translation of 3 squares left, 1 square up?

.... and ....

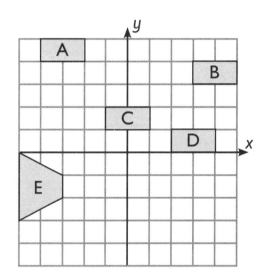

b) Which two rectangles show a translation of 1 square right, 3 squares up?

.... and ....

c) Reflect shape E in the y-axis.

4

Shape A is translated 4 squares right and 2 squares down to shape B.

Write down the co-ordinates of the vertices of shape B.

(................, ...) (................, ...)

(................, ...) (................, ...)

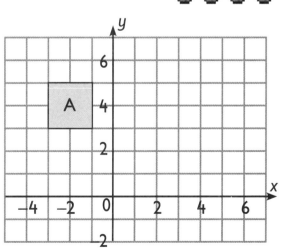

Above the platform where the chests sit is an item frame. Held there is an elytra; something Rosa has never seen before. She is unsure what this item is, or what it does. Taking the elytra from its frame, Rosa places it around her shoulders, thinking it may be a cape. It's only when she jumps off the ship and onto the End island that the elytra opens up and Rosa takes flight!

5

Shape A′ is the result of a translation of shape A.

Shape A was translated 8 squares left and 3 squares up to make shape A′.

Draw shape A on the grid.

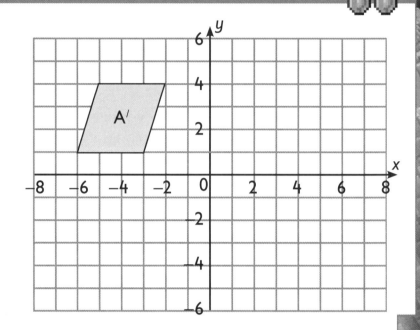

6

 a) Reflect shape A in the y-axis.

Label the new shape B.

b) Reflect shape B in the x-axis.

Label the new shape C.

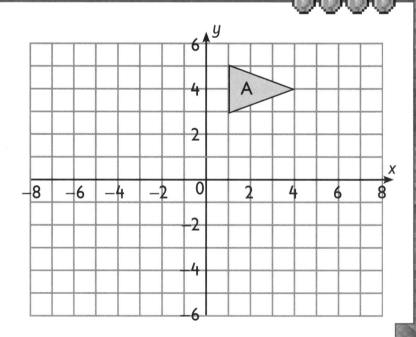

COLOUR IN HOW MANY EMERALDS YOU EARNED

ADVENTURE ROUND-UP

ROSA RISES

Rosa soars through the air. As she dives down, she gathers speed before blasting back up into the sky. This is something she has never experienced before. Gliding over the End island, Rosa cannot help but smile and think of the wonderful possibilities this item brings. Maybe she can use fireworks to fly faster, and further!

WHAT A VIEW!

Feeling daring, Rosa flies as high as possible. Looking down on the islands below, the chorus trees look tiny. She can also see the tower she climbed. Squinting at the gateway, she can see the gap which leads back home.

RETURN FLIGHT

Tipping suddenly forward, Rosa gains speed. She has seen everything she wants to see here and it's time to go home. Going faster than ever, Rosa angles her body perfectly and bursts through the gap in the gateway, disappearing from the End islands and back to where she fought the Ender Dragon. She will be home very soon!

HOME IS WHERE THE HEART IS

Back in the plains, Isaac is standing in his garden looking out at the trees. Some sheep climb over a small hill into view and begin grazing on the grass. This was the best place to build a house, he thinks. It has got everything a hero can want — there are animals, seeds and plenty of building materials.

A SPECTACULAR ARRIVAL

As Isaac turns towards the farm behind his house, he hears a whooshing noise in the distance. The noise steadily grows louder and he begins to worry. Isaac searches the sky for the source of the noise. Before he knows it, Rosa drops out of the air and onto the roof, grinning and waving! The wings fold up on her back and she climbs down to see her friend.

LOTS TO LOOK BACK ON

Rosa laughs as she sees the look of shock on Isaac's face. While he stands speechless, she explains about the elytra and the End islands. Once Isaac has taken it all in, they decide it's time to spend some quality time on familiar ground in the plains.

LINE GRAPHS

It is going to be another pleasant day in the plains so Rosa and Isaac will make the most of it.

I

The graph shows the temperature between 10 am and 10 pm yesterday.

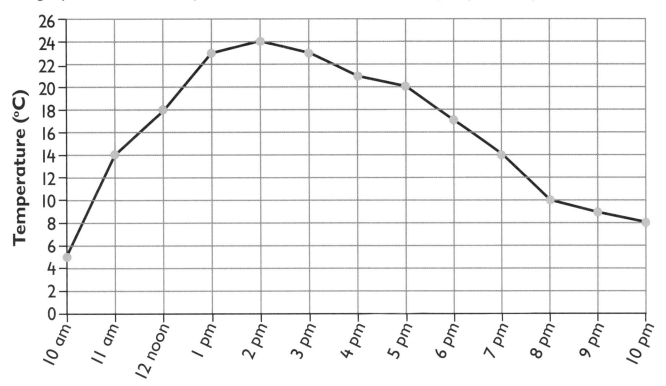

Time of day

a)   In what units was the temperature measured? ....

b)   What was the lowest temperature recorded? .... °C

c)   At what time was the highest temperature recorded? ....

d)   How often was the temperature recorded? ....

e)   How much colder was it at 9 pm than at 3 pm? .... °C

f)   At what times was the temperature 14°C? .... and ....

g)   At what time did the temperature start decreasing? ....

PIE CHARTS

Rosa reflects on the fight with the Ender Dragon. It was extremely dangerous, but they had a lot of fun using the tools and items they had created. Looking back, they are glad they spent so much time preparing.

1

Our heroes have 60 items in a chest.

The pie chart shows information about them.

Items in Chest

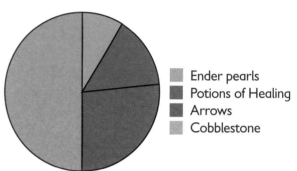

a) Which item do they have most of in the chest?

....

b) Which item do they have least of in the chest?

....

c) What fraction of the items in the chest are cobblestone?

d) How many cobblestone are in the chest? .... cobblestone

e) They have 5 Ender pearls and 16 arrows in the chest.

How many Potions of Healing do they have? .... Potions of Healing

---

Isaac recalls his long trek to the snowy tundra.

2

The pie chart shows the different types of food that Isaac ate on his journey.

Food Consumption

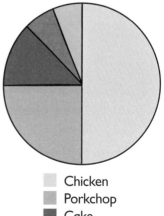

a) Isaac ate 16 chicken.

How many porkchops did he eat? .... porkchops

b) What total of food items is represented in the pie chart?

.... food items

c) What fraction of the food eaten was cake?

d) Estimate how many apples Isaac ate. .... apples

As the subject of food comes up, Isaac asks Rosa if she would help him to harvest some crops on his farm.

3

The table shows how many fields grow different crops on Isaac's farm.

Crop	Number of fields
Potatoes	3
Beetroot	1
Wheat	2
Total	6

Colour this pie chart to show the information.

Colour the key for the pie chart.

Key: Potatoes ☐

Beetroot ☐

Wheat ☐

Isaac's Fields

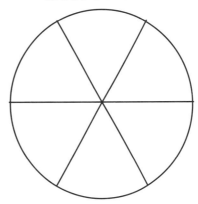

4

 The table gives some information about the animals on Isaac's farm.

Draw a pie chart to represent this information.

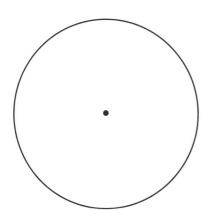

	Number of animals
Cow	4
Sheep	2
Horse	1
Llama	1
Total	8

Make a key and a title for your pie chart.

THE MEAN

To find the mean of these numbers:	3, 6, 8, 11
Add the four numbers:	3 + 6 + 8 + 11 = 28
Divide the total by how many numbers you added:	28 ÷ 4 = 7

Farming isn't just about physical work. Isaac and Rosa need to use their heads to do some calculations. Can you help them out?

1

Here are three batches of carrots:

Work out the mean number of carrots in a batch. .... carrots

2

Here are the masses of four hay bales:

Work out the mean mass of the hay bales. .... kg

3

 The mean of these four numbers is 14.

What is the missing number? ....

COLOUR IN HOW MANY EMERALDS YOU EARNED

ADVENTURE ROUND-UP

A WELL-DESERVED BREAK

It's now time to relax. The two friends stand together, looking out over the farm and reflecting on everything they achieved.

FOREVER FRIENDS

Rosa and Isaac have seen so many marvellous things both in the Overworld and in other dimensions, overcoming all the challenges they faced. They even passed the ultimate test by conquering the mighty Ender Dragon. They couldn't have done it without each other. The special tools and jewels are nice, but the most valuable thing they have is their friendship. That's something to really treasure.

ANSWERS

Pages 5–7

1. a) 3,611,200 [1 emerald]

   b) two million, eight hundred and nine thousand [1 emerald]

2.

Blocks travelled	Rounded to nearest 1,000	Rounded to nearest 10,000	Rounded to nearest 1,000,000
4,092,781	4,093,000	4,090,000	4,000,000
6,935,092	6,935,000	6,940,000	7,000,000
8,249,711	8,250,000	8,250,000	8,000,000

[1 emerald for each correct row]

3. a) < b) >

   c) < d) < [1 emerald each]

4. a) three hundredths (0.03) [1 emerald]

   b) thirty thousand (30,000) [1 emerald]

   c) three thousandths (0.003) [1 emerald]

5. 1,491,226 coloured in [1 emerald]

6. 326.2, 32.6, 3.662, 3.262, 3.26 [1 emerald]

7.

	Rounded to 1 decimal place	Rounded to the nearest whole number
4.25	4.3	4
7.41	7.4	7
9.82	9.8	10

[1 emerald for each correct row]

8. 0.099 circled [1 emerald]

Pages 8–9

1. a) 4.5°C [1 emerald]

   b) −3.5°C [1 emerald]

2. a) A: −4°C   B: −6°C   C: −20°C [1 emerald each]

   b) C [1 emerald]

3. −2°C [1 emerald]

4. a) −3 b) −7

   c) −3 d) 2

   e) 5 f) 2 [1 emerald each]

5. Line drawn to join −7 and 8 [1 emerald]

Pages 10–11

1. a) 295,843 [1 emerald]

   b) 371,607 [1 emerald]

   c) 3,528,910 [1 emerald]

   d) 7,990,190 [1 emerald]

2. a) 77 b) 9,999

   c) 120 d) 40

   e) 149 f) 229

   g) 7,034 h) 1,745 [1 emerald each]

3. a) £4.69 [1 emerald]

   b) £5.31 [1 emerald]

4. a) 470 b) 660

   c) 90,891 d) 57,200

   e) 5,000 f) 3,699 [1 emerald each]

5. 61 + 89 = 150; 299 − 150 [1 emerald]
   = 149 coal [1 emerald]

Pages 12–13

1. a) 35,610 b) 453.72

   c) 714.2 d) 640,345 [1 emerald each]

2. a) 376,148.6 b) 3,290.89

   c) 535.24 d) 288.14 [1 emerald each]

3. a) 11,150 b) 9,626,798

   c) 487.02 d) 4,841.54

   e) 1,942.4 f) 27.753 [1 emerald each]

4. Box in bottom row: 47,442 [1 emerald]
   Box in middle row: 47,705 [1 emerald]
   Box in top row: 98,892 [1 emerald]

5. a) 54,621 − 23,831 = 30,790 [1 emerald]

   b) 29,760 + 63,416 = 93,176 [1 emerald]

Pages 14–15

1. a) 6,000 + 3,000 − 5,000 = 4,000 [1 emerald]

   b) 7,000 + 1,000 + 4,000 = 12,000 [1 emerald]

   c) 9,000 − 3,000 − 3,000 = 3,000 [1 emerald]

   d) 7,000 − 5,000 − 1,000 = 1,000 [1 emerald]

2. Any suitable answers. **Examples:**

   a) 77,654 − 23,782 = 53,872 [1 emerald]

   b) 23,878 + 37,058 = 60,936 [1 emerald]

   c) 51,154 − 18,543 = 32,611 [1 emerald]

   d) 45,462 + 47,675 = 93,137 [1 emerald]

3. Any suitable answer. **Example:**
   11,128 + 6,902 = 18,030 (not 17,826) [1 emerald]

4. a) 700,000 + 500,000 = 1,200,000 [1 emerald]

   b) 1,500,000 − 400,000 = 1,100,000 [1 emerald]

   c) 3,400,000 + 900,000 + 200,000 = 4,500,000 [1 emerald]

   d) 2,600,000 − 600,000 − 400,000 = 1,600,000 [1 emerald]

5. 43 − 20 + 4 − 9 [1 emerald]
   = 18 [1 emerald]

Pages 16–18

1. 20,000 − 12,250 − 2,750 − 1,800 [1 emerald]
   = 3,200 [1 emerald]

2. 4,845 + 5,177 = 10,022 [1 emerald]
   Yes, he wins a prize [1 emerald]

3. 11,299 − 7,496 [1 emerald]
   = 3,803 [1 emerald]

4. 14,814 − 1,893 − 735 [1 emerald]
   = 12,186 [1 emerald]

5. a) 107.22 seconds [1 emerald]

b) 12.34 seconds [1 emerald]
c) 2 and 3 [1 emerald]
d) 7.63 seconds [1 emerald]
6 a) Plains [1 emerald]
b) 28,796,742 [1 emerald]
7 258,421 − 100,000 − 96,775 [1 emerald]
= 61,646 [1 emerald]
8 £10,000 − £500 − £1,250 − £59.75 − £312.40 − £5,226.04

[1 emerald]
= £2,651.81 [1 emerald]

Pages 21–23

1 a) Circled: 60, 180, 300 [1 emerald]
b) Circled: 50, 100, 350 [1 emerald]
2 a) 6   **b)** 60   **c)** 100 [1 emerald each]
3 a) 1, 2, 4 and 8 [1 emerald]
b) 96 [1 emerald]
4 a) >   **b)** =   **c)** > [1 emerald each]
5 41 is a prime number [1 emerald]
(other prime numbers that can be made are 13, 23, 31, 43, 53, 61)
56 is a common multiple of 8 and 28 [1 emerald]
32 is a common factor of 64 and 96 [1 emerald]
(another common factor that can be made is 16)
6 a) 1,056   **b)** 77 [1 emerald each]
7 Possible answers: 13 (or 31); 17 (or 71); 37 (or 73); 79 (or 97) [1 emerald]
8 Possible answers: 3 + 1 = 4; 7 + 9 = 16

[1 emerald for suitable numbers on the left; 1 emerald for a suitable total on the right]

Pages 24–25

1 a) 24.5   **b)** 510
c) 1.56   **d)** 1,600
e) 90   **f)** 1,500
g) 60   **h)** 400 [1 emerald each]
2 a) 0.51   **b)** 598
c) 2.619   **d)** 90.63
e) 0.207   **f)** 604
g) 0.431   **h)** 7.8 [1 emerald each]
3 a) 5.6   **b)** 13.56
c) 340   **d)** 7,291
e) 0.478   **f)** 24,730
4 3.5 kg [1 emerald]
5 a) 4.2   **b)** 3.6
c) 2.4   **d)** 0.2
e) 0.72   **f)** 0.56
g) 12   **h)** 9 [1 emerald each]

Pages 26–27

1 1,008 books [1 emerald]
2

×	400	70	3
30	12,000	2,100	90
6	2,400	420	18

[1 emerald]

Total: 17,028 [1 emerald]
3 a) 16,832 [1 emerald]
b) 38,428 [1 emerald]
c) 1,086.24 [1 emerald]
4 £183 [1 emerald]
5 506,678 [1 emerald]
6 a)

```
    5 [4] 6  3
  ×     [3] 6
  ─────────────
    3  2  7  7  8
  1  6  3  8 [9] 0
  ─────────────
  1  9  6  6  6  8
```
[1 emerald]

b)

```
  3  1 [6][7]
  ×      2  7
  ─────────────
  2  2  1  6 [9]
  6  3  3  4  0
  ─────────────
  8  5  5  0  9
```
[1 emerald]

Pages 28–29

1 a) 17   **b)** 18
c) 44   **d)** 12
e) 170   **f)** 14.3 [1 emerald each]
2 27p [1 emerald]
3 a) 890   **b)** 421
c) 205   **d)** 0.19 [1 emerald each]
4 16 storage slots [1 emerald]
5 70 mossy stone bricks [1 emerald]

Pages 30–31

1 a) 15   **b)** 8 [1 emerald each]
2 a) 38.6   **b)** 41.25
c) 15.5   **d)** 105.75
e) 128.125   **f)** 24.15 [1 emerald each]
3 7,052 [1 emerald]
4 19 [1 emerald]
5 a) $15\frac{1}{2}$ **b)** $36\frac{1}{3}$
c) $16\frac{1}{4}$ **d)** $41\frac{9}{14}$
e) $63\frac{9}{22}$ **f)** $122\frac{1}{5}$ [1 emerald each]

Pages 32–33

1 9 × 1,368 = 12,312 [1 emerald]
12,312 × 19 = 233,928 passengers [1 emerald]
2 35 × 64 = 2,240 [1 emerald]
800 × 2 = 1,600 [1 emerald]
2,240 − 1,600 = 640 carrot seeds [1 emerald]
3 a) 2,744   **b)** 14.75
c) 18   **d)** 3,205 [1 emerald each]
4 a) £166.50 + £52.50 [1 emerald]
= £219 [1 emerald]
b) £357 − £161 = £196 [1 emerald]
£196 ÷ £3.50 = 56 acacia trees [1 emerald]
c) 9,600 ÷ 400 [1 emerald]
= 24 lorry loads [1 emerald]
5 50 × 12 = 600 [1 emerald]
174 + 56 = 230 [1 emerald]
600 − 230 = 370 trees [1 emerald]

Page 34

1. a) 30   b) -2
   c) 40   d) 180
   e) 36   f) 20   [1 emerald each]
2. a) 110   b) 8
   c) 4,000   d) 51,100
   e) 21   f) 36   [1 emerald each]
3. a) 2   b) 4   c) 4   [1 emerald each]

Page 37

1. $\frac{2}{3} = \frac{4}{6} = \frac{6}{9} = \frac{8}{12} = \frac{10}{15} = \frac{12}{18}$ [1 emerald]
2. a) $\frac{3}{5}$ b) $\frac{3}{7}$
   c) $\frac{2}{9}$ d) $\frac{3}{4}$
   e) $\frac{1}{3}$ f) $\frac{1}{3}$
   g) $\frac{5}{3}$ h) $\frac{3}{25}$ [1 emerald each]
3. a) $\frac{2}{3}$ $\frac{7}{10}$ $\frac{11}{15}$ $\frac{4}{5}$ [1 emerald]
   b) $\frac{3}{5}$ $\frac{5}{8}$ $\frac{13}{20}$ $\frac{27}{40}$ [1 emerald]
   c) $5\frac{1}{5}$ $5\frac{3}{10}$ $5\frac{1}{3}$ $5\frac{5}{6}$ [1 emerald]

Pages 38–39

1. a) $\frac{3}{10}$ b) $\frac{2}{8}$ or $\frac{1}{4}$ c) $\frac{2}{9}$ [1 emerald each]
2. a) 6 rectangles shaded black; 9 rectangles shaded
      green; 4 rectangles shaded yellow   [1 emerald]
   b) Black: 6; Green: 9; Yellow: 4   [1 emerald]
   c) $\frac{5}{24}$ [1 emerald]
3. a) i) 4   ii) 3   iii) 5
      [1 emerald each]
   b) i) 3   ii) 8   iii) 20
      [1 emerald each]
4. 42 gold ingots   [1 emerald]
5. a) $\frac{4}{5}$ b) 6 blocks   c) 30 blocks
      [1 emerald each]

Pages 40–41

1. a) $\frac{3}{6}$ (or $\frac{1}{2}$)   b) $\frac{2}{8}$ (or $\frac{1}{4}$) $+ \frac{1}{8} = \frac{3}{8}$
   c) $\frac{4}{8}$ (or $\frac{1}{2}$) $+ \frac{3}{8} = \frac{7}{8}$ d) $\frac{6}{12}$ (or $\frac{1}{2}$) $- \frac{3}{12}$ (or $\frac{1}{4}$) $= \frac{3}{12}$ (or $\frac{1}{4}$)
      [1 emerald each]
2. a) $\frac{5}{12}$ b) $\frac{13}{12}$ (or $1\frac{1}{12}$)
   c) $\frac{19}{21}$ d) $\frac{4}{15}$
   e) $\frac{6}{12}$ (or $\frac{1}{2}$)   f) $\frac{3}{20}$ [1 emerald each]
3. $\frac{7}{15}$ [1 emerald]
4. a) $4\frac{7}{8}$ b) 5
   c) $\frac{5}{12}$ d) $2\frac{1}{30}$ [1 emerald each]
5. a) 4   b) 10   c) 3   [1 emerald each]

Pages 42–43

1. a) $\frac{1}{40}$ b) $\frac{1}{100}$
   c) $\frac{2}{18}$ (or $\frac{1}{9}$)   d) $\frac{9}{20}$ [1 emerald each]
2. a) $\frac{1}{4}$ b) $\frac{1}{12}$
   c) $\frac{1}{10}$ d) $\frac{1}{30}$
   e) $\frac{3}{32}$ f) $\frac{5}{30}$ (or $\frac{1}{6}$)   [1 emerald each]
3. Any suitable answers. **Examples:**
   a) $\frac{1}{4}$ b) $\frac{1}{16}$
   c) $\frac{1}{3}$ d) 5
   e) $\frac{1}{4} \times \frac{1}{3} = \frac{1}{6} \times \frac{1}{2}$ f) $\frac{5}{4} \div 2 = \frac{1}{4} \times \frac{5}{2}$ [1 emerald each]

4. $\frac{4}{5} \div 3$ [1 emerald]
   $= \frac{4}{15}$ [1 emerald]
5. $\frac{3}{4} \times \frac{1}{2}$ [1 emerald]
   $= \frac{3}{8}$ [1 emerald]

Pages 44–45

1. a) 0.5   b) 0.75
   c) 0.2   d) 0.1
   e) 0.625   f) 0.35   [1 emerald each]
2. a) $\frac{9}{100}$ b) $\frac{71}{100}$
   c) $\frac{437}{1,000}$ d) $\frac{103}{1,000}$
   e) $\frac{6}{25}$ f) $\frac{3}{5}$
   g) $\frac{7}{20}$ h) $\frac{3}{8}$
   i) $\frac{9}{20}$ [1 emerald each]
3. a) 0.2   0.5   0.9   [1 emerald]
   b) 0.05   0.3   0.45   [1 emerald]
   c) 0.01   0.03   0.045   [1 emerald]
4.
   [1 emerald each]
5. 0.04   0.3 $\frac{1}{3}$ $\frac{2}{5}$ 0.76   [1 emerald]

Pages 46–47

1. a) 20 arrows   [1 emerald]
   b) 20 melon slices   [1 emerald]
   c) 75 torches   [1 emerald]
2. a) $\frac{9}{10} = 0.9$ b) $\frac{35}{100}$ (or $\frac{7}{20}$) $= 0.35$
   c) $\frac{75}{100}$ (or $\frac{3}{4}$) $= 0.75$
   d) $\frac{40}{100}$ (or $\frac{4}{10}$ or $\frac{2}{5}$) $= 0.4$ [1 emerald each]
3. 8 emeralds   [1 emerald]
4. Rosa (65% compared to Isaac's 60%)   [1 emerald]
5. Diamond boots: £19.20   [1 emerald]
   Diamond leggings: £40.80   [1 emerald]
   Diamond helmet: £22   [1 emerald]

Pages 48–50

1. Any 12 triangles of the shape shaded   [1 emerald]
2. 0.08   60% $\frac{3}{4}$ $\frac{8}{10}$ [1 emerald]
3. a) Creepers: 40   Skeletons: 100
      Zombies: 240   Phantoms: 20   [1 emerald each]
   b) 60   [1 emerald]
4. $\frac{4}{5}$ $\frac{7}{10}$ $\frac{2}{3}$ $\frac{9}{15}$ [1 emerald]
5. a) $\frac{3}{4}$ [1 emerald]
   b) 5 potatoes   [1 emerald]
   c) 20 potatoes   [1 emerald]
6. a) $\frac{1}{6}$ b) $\frac{1}{2}$
   c) $\frac{1}{4}$ d) $\frac{1}{4}$ [1 emerald each]
7. Cube in bottom row: $1\frac{1}{4}$ [1 emerald]
   Cube in middle row: $3\frac{3}{4}$ [1 emerald]
   Cube in top row: $7\frac{1}{4}$ [1 emerald]
8. a) 60   b) 39   [1 emerald each]

Page 53

1 3 bottles coloured purple and 9 bottles coloured pink [1 emerald]

2 a) 1 : 3   b) 1 : 2
   c) 1 : 8   d) 1 : 5   [1 emerald each]

3 27 Arrows of Weakness [1 emerald]

Pages 54–55

1 a) 3.6 energy   b) 14.4 energy   [1 emerald each]

2 26 m [1 emerald]

3 a) 7 cm   b) 28 cm   [1 emerald each]

4 7 times longer [1 emerald each]

5 a) £2.40   b) 3 kg   [1 emerald each]

Pages 56–57

1 Table completed from top: 6; 15; 30 [1 emerald each]

2 Isaac: 40   Rosa: 10 [1 emerald each]

3 120 emeralds [1 emerald]

4 Discussing strategy: 60 seconds [1 emerald]
Eating: 40 seconds [1 emerald]
Checking weapons: 20 seconds [1 emerald]

5 360 g [1 emerald]

Pages 58–59

1 a) $x = 6$ b) $y = 23$ c) $n = 7$ [1 emerald each]

2 Arrow = 29; Crystal = 1 [1 emerald each]

3 30 seconds [1 emerald]

4 $4x - 3 = 21$ [1 emerald]

5 Any suitable answers. **Examples:**
$a = 7$ and $b = 1$ [1 emerald]
$a = 8$ and $b = 2$ [1 emerald]
$a = 9$ and $b = 3$ [1 emerald]

Page 60

1 a) −3   −12   −21 [1 emerald]
  b) 48.5   57   65.5 [1 emerald]

2 a) Add 4 [1 emerald]
  b) Subtract 8 [1 emerald]

3 a) 27   54 [1 emerald]
  b) 63   70   77 [1 emerald]

Page 63

1 Perimeter = 10 km   Area = 6 km² [1 emerald each]

2 a) 40 m   b) 71 m² [1 emerald each]

3 100 m² (a square 10 m by 10 m) [1 emerald]

Pages 64–65

1 96 cm² [1 emerald]

2 a) 30 cm²   b) 75 cm² [1 emerald each]

3 Head: 25 cm² [1 emerald]
Middle part of body: 8 cm² + 8 cm² + 16 cm² = 32 cm² [1 emerald]
Legs: 13 cm² + 13 cm² = 26 cm² [1 emerald]
Arms: 17 cm² + 17 cm² = 34 cm² [1 emerald]
Total: 117 cm² [1 emerald]

4 base = 30 cm [1 emerald]

Pages 66–67

1 a) 1,920 m³   b) 4,500 m³ [1 emerald each]

2 96 − 38 [1 emerald]
= 58 [1 emerald]

3 216 cm³ [1 emerald]

4 2 cm [1 emerald]

5 Any three values that multiply to make 240. **Example:**
Length: 15 cm   Width: 4 cm   Height: 4 cm [1 emerald each]

Page 68

1 a) 1 pint [1 emerald]
  b) 35 centimetres [1 emerald]
  c) 1 mile [1 emerald]

2 a) 6,500 g   b) 79 mm
  c) 935 cm   d) 3.975 km
  e) 10.825 l   f) 8.025 kg [1 emerald each]

3 Any answer from 185 to 190 miles [1 emerald]

Pages 71–73

1 Second circle should be ticked [1 emerald]

2 6 m [1 emerald]

3 a) D   b) A, B, C, D
  c) B, D   d) B [1 emerald each]

4

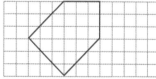

[1 emerald for each correct side]

5 23 cm [1 emerald]

6 50 cm [1 emerald]

7 Any suitable answer. **Example:**

[1 emerald for a pentagon; 1 emerald for three right angles]

8 The circles touch the perimeter 12 times, so
120 ÷ 12 = 10 cm diameter per circle [1 emerald]
Radius of one circle: 5 cm [1 emerald]

Pages 74–75

1 a) Triangular prism and square-based pyramid [1 emerald]
  b) Triangular prism and cuboid [1 emerald]

2 a) Cube [1 emerald]
  b) Cuboid [1 emerald]

3 C [1 emerald]

4 Faces: 7 [1 emerald]
Edges: 15 [1 emerald]
Vertices: 10 [1 emerald]

5 Cuboid [1 emerald]

Pages 76–77

1 **a)** $x = 25°$ **b)** $x = 45°$
   c) $x = 29°$ **d)** $x = 263°$ [1 emerald each]

2   $x = 107°$ [1 emerald]

3   $y = 34°$ [1 emerald]

4   $180 - 72 = 108$ [1 emerald]
   All angles in a regular pentagon are equal so $x = 108°$
                            [1 emerald]

5 **a)** $x = 72°$ $y = 108°$ [1 emerald each]
   b) $x = 68°$ $y = 67°$ [1 emerald each]

Pages 78–79

1 **a)** (1, 1)   (−3, 0)   (−2, −4)   [1 emerald each]
   b) (2, −3)   [1 emerald]

2 **a)**

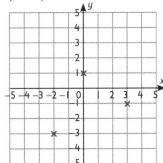

                          [1 emerald]
   b) (1, −5) or (5, 3) or (−5, −1)   [1 emerald]

3 (4, −3)   [1 emerald]

4 **a)**

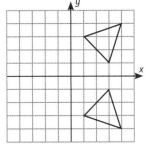

                          [1 emerald]
   b) Points joined to form a trapezium   [1 emerald]
   c) (−2.5, 3)   [1 emerald]

5 (1, −2)   [1 emerald]

Pages 80–82

1 **10** squares right and **2** squares **up**   [1 emerald each]

2

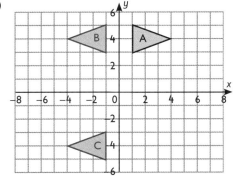

                 [1 emerald for each correct side]

3 **a)** D and C   [1 emerald]
   b) D and B   [1 emerald]

c)

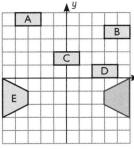

                          [1 emerald]

4 (3, 3)   (3, 1)   (1, 3)   (1, 1)   [1 emerald each]

5

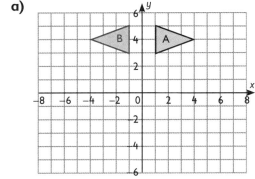

[2 emeralds if fully correct; 1 emerald if one error only]

6 **a)**

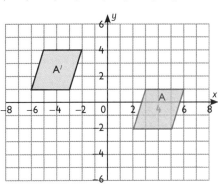

[2 emeralds if fully correct; 1 emerald if one error only]

b)

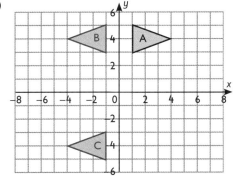

[2 emeralds if fully correct; 1 emerald if one error only]

Page 85

1 **a)** degrees Celsius (or °C)   [1 emerald]
   b) 5°C   [1 emerald]
   c) 2 pm   [1 emerald]
   d) Every hour   [1 emerald]
   e) 14°C   [1 emerald]
   f) 11 am and 7 pm   [1 emerald]
   g) 2 pm   [1 emerald]

Pages 86–87

1 a) Cobblestone   [1 emerald]
   b) Ender pearls   [1 emerald]
   c) $\frac{1}{2}$ [1 emerald]
   d) 30 cobblestone   [1 emerald]
   e) 9 Potions of Healing   [1 emerald]

2 a) 8 porkchops   [1 emerald]
   b) 32 food items   [1 emerald]
   c) $\frac{1}{8}$ [1 emerald]
   d) 2 apples   [1 emerald]

3 Any suitable answer. **Example:**

Isaac's Fields

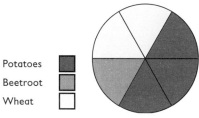

Key: Potatoes
       Beetroot
       Wheat

[2 emeralds for crops correctly represented on pie chart; 1 emerald for correct key]

4 Any suitable answer. **Example:**

Isaac's Animals

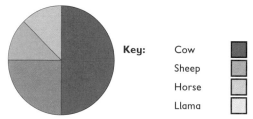

Key: Cow
       Sheep
       Horse
       Llama

[2 emeralds for animals correctly represented on pie chart; 1 emerald for key; 1 emerald for title]

Page 88

1 7 carrots   [1 emerald]
2 16 kg   [1 emerald]
3 10   [1 emerald]

TRADE IN YOUR EMERALDS!

Well done for helping Rosa and Isaac to complete their quest! Throughout this book, you earned emeralds for your hard work. This merchant is now waiting for you to spend your gems with them. Pretend you will build a large and grand house in your biome of choice. This will be your base; a place to sleep, craft and prepare for adventure. Which materials will you buy to build this home?

If you have enough emeralds, you could buy more than one set of some items.

Count all your emeralds and write the total in this box:

HMMM?

SHOP INVENTORY

64 X CALCITE: 15 EMERALDS

32 X AMETHYST: 12 EMERALDS

64 X POLISHED ANDESITE: 8 EMERALDS

64 X POLISHED GRANITE: 6 EMERALDS

32 X BLOCK OF COPPER: 30 EMERALDS

64 X RED GLAZED TERRACOTTA: 24 EMERALDS

64 X BLOCK OF TERRACOTTA: 16 EMERALDS

32 X POLISHED DEEPSLATE BLOCK: 12 EMERALDS

64 X BLUE GLAZED TERRACOTTA: 20 EMERALDS

64 X STONE BLOCK: 5 EMERALDS

64 X YELLOW GLAZED TERRACOTTA: 25 EMERALDS

64 X STONE BRICK STAIRS: 10 EMERALDS

64 X CHISELLED SANDSTONE BLOCK: 20 EMERALDS

64 X QUARTZ PILLAR: 50 EMERALDS

64 X QUARTZ BLOCK: 40 EMERALDS

That's a lot of emeralds. Well done! Remember, just like real money, you don't need to spend it all. Sometimes it's good to save up.